CONTENTS

PROLOGUE

THE FLOATING HOUSE

THE FIRST TIME Lilian Roth saw the haunted house it was floating down Main Street, and if she'd known the trouble it would cause them, she never would have pointed it out to her best friend.

She and Ben Laramie were winding their way through *Infinite Zombie*'s abandoned hospital level for the hundredth time when her living room suddenly went dark and the temperature dropped, causing goosebumps to spring up on her arms. As a shadow passed over the TV screen, she stood abruptly, spilling her can of cream soda all over the carpet. Paying it no mind —her mother would be upset, Dad would bail her out—Lil hurried to the window to look.

The Roths lived on Main Street, above the bulk food store and across from one of two antique shops in Duck Falls. She squinted through the fly-specked window screen, expecting to see the crumbling façade and flat tar roof of Green's Antiques. Instead she found herself staring at a brick wall crawling with flat green leaves, so close she could almost reach out and touch it if she hung precariously out the window. She smelled its cool, rotting brick and the crisp scent of the leaves, confirming it wasn't a hallucination. Two deep-set windows moved by, and through their warped glass she saw more antique furniture than Mr. and Mrs. Green might sell in a year from their shop: a large wooden desk and shelves full of books, a massive globe, an armchair under a drop cloth, a lion's head mounted on the far wall near a stone fireplace and an open doorway.

Why would they move it with all the furniture still inside? she thought. For some reason, this question troubled her most of all.

"Holy crap, Ben! Do you see that?"

"I saw that zombie totally rip your head off," Ben said over the headset from his own house. "I was just about to blow its brains out, why'd you pause the game?"

"Go to your window."

"Let's just finish this level," he whined. "I've only got an hour 'til Mom's home and I still gotta mow the back lawn."

"*Go to the window,*" she said again.

Through the doorway inside the moving house, an enormous shadow swept across the wall. For a moment it appeared to have several arms, like a human octopus. Either that or it belonged to more than one person, a group of —*creatures*—people huddled together in the hall, like stowaways. Before she could decide which it was, the shadow was gone, disappearing into the patterned wallpaper. Whether it had been a trick of the light or a brief hallucination caused by the shock of watching a house literally float past her window, her heart thrummed unpleasantly fast.

As the black door inside the house began to creak closed, she lowered the blinds and backed away, struck with a sudden certainty she'd just made a terrible mistake, like stepping in fresh-laid concrete or calling someone she loved an unforgivable name.

Inside the moving house, the black door clicked shut.

Frantic, Lil opened her mouth to prevent Ben from picking up his binoculars. But her words caught in her throat. She stood there gawping like a dying fish, watching the darkness pass her apartment through the blinds.

Ben let out an annoyed huff and got up from the floor, knowing if he didn't at least humor her he'd be "in the doghouse," as his dad sometimes said regarding his mother. His legs and butt numb from sitting cross-legged on the rug while the two of them played their favorite survival horror game, Ben hobbled to his bedroom window to look. "Where am I looking?"

Lil didn't answer for a moment. "Never mind," she said finally. "It's nothing."

Ben grabbed the small pair of folding binoculars off his desk. "You sounded pretty tweaked out for 'nothing.'"

He peered through the eyepieces, knowing he'd be able to see whatever Lil had, despite the distance. From his bedroom window, way up on the hill old folks still called the "Duck Bill," he could see almost the entire town with the binoculars he'd gotten a few Christmases back. He and Lil had tested them, with Ben warm inside his house and Lil braving the cold, standing in the middle of the street out front of her apartment, waving madly and shivering. When he pulled the focus now, he wasn't prepared for what he would see, and he wasn't sure he could believe his eyes. He blinked hard and peered through them again only to be visited by the same impossible sight.

A house hovered down Main Street like an apparition, like something out of a nightmare. The skin on his neck and cheeks prickled electric with disbelief and—despite his self-professed desensitivity to All Things Horror—even fear. He lowered the binoculars a moment, his palms slick with sweat. The hair on his arms stood on end. In the novels he enjoyed, the writers often called this phenomenon "horripilation." He'd always liked the ring of the word but had never experienced it quite as he was now.

When he zoomed out, he was relieved to discover the house wasn't floating down the street on its own. Dozens of large wheels rolled beneath it, but the cab or tractor or whatever carried the house was hidden from view behind its covered porch.

He remembered workers had been doing something with the power and telephone lines downtown on his way home from dinner at Grandma Laramie's house the other night, and now he knew the reason. They'd been making space to move the house through town.

"That can't be real," he muttered into the headset, pulled taut from the console. He'd spoken it aloud but not to Lil, only to convince himself.

"It's real," she said. "There was furniture inside and..."

The sun had brightened her living room again, warming her goosebumped arms. Now that it was gone, she felt a surge of bravery. She drew up the blinds, pulled the bent screen off its frame, and stuck her head out the window. Several kids she recognized chased the house on their bikes, cheering and whistling and popping wheelies. Adults and children had stopped on both sides as if to watch a parade, shading their eyes to get a better look.

"And what?" Ben asked.

"It was nothing," she said, more to assuage her own fears than to answer Ben. She shook her head in amusement, chuckling at her initial fear. It was just a big old ugly house, nothing all that scary about it. If it hadn't rolled by her living room windows, if she'd seen it at the end of a cul-de-sac or at the top of the Duck Bill, she could imagine an old lady in a sunhat picking weeds from flowerbeds out front, or a middle-aged guy with his gut hanging over his jeans drinking a beer and pushing a mower across its lawn.

Ben tailed the house with the binoculars. It looked straight out of the old horror movies he loved: three stories high with gabled rooftops like jagged teeth, veins of dark green ivy crawling over crumbled bricks the color of dried blood, shutters on the windows. He thought it even looked kind of familiar, but in his excitement, he couldn't quite place it. Maybe it had been used in a horror movie.

A haunted house, he thought. *A real haunted house rolling up Main Street. Wow!*

"It's gone," Lil said half the town away, relief in her voice.

With the house out of sight behind the Post Office, the people in the street resumed their business, returning to stores and carrying on about their day as if nothing out of the ordinary had just happened—though Lil suspected none of them had seen anything so interesting in their lives, especially not right here in the boring old town of Duck Falls, Maryland.

Nothing to be afraid of, she thought. *Just a house—a big, ugly old house.*

She thought of the shadow she'd seen in the hall, but she quickly pushed the image away. What good would playing that game do, besides creeping herself out for no reason?

"Well, why don't you go outside?" Ben suggested. To look at the house, he meant.

"Headset," she said. "Duh." But it was really just an excuse. She didn't want to be anywhere near that house, haunted or not. And she hoped wherever it was going was very far away from here.

From high on the Duck Bill, Ben could still see the vaguely familiar old house continue through town. He thought it might be heading toward the old fairgrounds, but he supposed it could be headed for Burt Bucklebee's farm. The real estate company Ben's mom worked for had brokered the deal between Bucklebee and some billion-dollar conglomerate called the Hedgewood Foundation. But that land had been rezoned a few years ago for a theme park, so Ben didn't think it would be going there.

Lil said, "Where do you think it came from?"

"No idea." All he knew for sure was that it wasn't from anywhere around Duck Falls. Having been dragged along on several house tours with his mom, he recognized the style as Victorian, which meant it was at least a hundred years old, possibly more. Older than the Laramie's house by at least fifty years. Even their hometown—often called "Duck Farts" by local wiseasses like the two of them—wasn't as old as the house now rolling away from the main drag. Duck Falls had been founded in 1919.

"What d'you think they're gonna do with it?" Lil asked after some time.

"Looks like they're moving it somewhere."

"Well no shit, Sherlock."

Slightly aggravated, Ben said, "Somebody probably bought it, now they're moving it to their property. My mom says they do that sometimes when the government or some company owns the land the house is on."

"Huh," Lil said. "Well, I guess she oughta know."

"Yup. It's called expropriation. I heard they flooded this one town in Canada for a dam, but you can still see the church steeple over the top of the lake, which I thought was pretty cool. Anyways, a lot of the houses got moved up the hill to make this new town..."

He trailed off, only talking to fill in the silence. As the house vanished behind First Methodist, leaving only its peaks visible above the steeple, he'd grown acutely aware of being alone in his own home. Even though it hadn't scared him for many years, the thought troubled him now. And when the house reemerged from behind the church, movement in a second-floor window caught his eye. The focus shifted, blurring what he thought he'd seen before his mind could thoroughly process it.

The hairs on his forearms were stiff as bristles. *Horripilation*, he thought.

He twisted the focus wheel, desperate to keep the retreating house in sight, assuring himself there was nothing to be afraid of: it was a strange lamp, a misplaced coat rack, a storefront mannequin—anything other than what he thought he'd seen.

Because what he thought was impossible.

Once he'd managed to get the binoculars focused again, the man he'd seen in the window was no longer there. The dirty lace curtains fluttered, but the motion of the house or a light breeze blowing in the opened windows could have easily caused it.

As he thought this, the man stepped back into the window, menacing dark eyes staring directly at him with telescopic accuracy. It was impossible for the man to see him with the distance between them, Ben knew that. But the thought of it startled him enough to drop the binoculars. While he fumbled with them, hung from the strap around his neck, he heard Lil planning their route through *Infinite Zombie*'s abandoned hospital level, as if she'd already forgotten about the actual haunted house that had just rolled past her apartment.

He settled the binoculars against the bridge of his nose, and found the man still in the window, still staring at him. Dressed in a button-down gray cardigan, he had wild dark hair, a thin mustache above red lips, and heavy bags under his piercing brown eyes. He looked like he hadn't slept in decades.

Ben knew it couldn't be him. It wasn't possible.

But he had to be sure.

Reluctantly, he set the binoculars down and grabbed the book he'd been reading off his bedside table. In that glossy red horror font particular to late-'70s and early '80s horror novels was the title, *THE HOUSE FEEDS*. Beneath was a haunted house very similar to the one in the street, although this had been painted in the classic '80s "horror boom" style with exaggerated features. Under the author's name was the tagline: *A novel of skin-prickling terror, soon to be a major motion picture!*

Heart drumming disconcertingly fast, Ben flipped the book over.

The man from the window stared out from the back cover in stark black and white. It was Rex Garrote—impossible and yet undoubtable—the writer *People* magazine had once called "the Most Terrifying Man in the World." The author of multiple best-selling horror novels, and creator and host of the short-lived TV series, *Rex Garrote's Ghost World*. Ben owned every single one of Garrote's books, all the comics, even a handful of screenplays.

But it couldn't possibly be Rex Garrote because for almost two decades Rex Garrote had been—

"*Dead*," Ben croaked.

"What?"

"I think..." He trailed off again. Forming the words had become difficult, and his tongue felt fat in his mouth, his speech slurring. "I think I saw..."

"What, Ben?"

He shook his head to clear the fog from his brain. Blood pounded thickly in his ears and his chest felt tight, as if a heavy weight were pressing on his ribs. His reflection on the television screen wavered and his bedroom grew dim. It seemed as though night had fallen outside but it was only three in the afternoon.

"What?" Lil asked. Her voice was far away, coming to him from the end of a long, dark tunnel, flashing warning red to the rapid beat of his heart. "Saw *what*, Ben?"

He staggered back, blindly reaching for his desk, for a chair to collapse in, for something to potentially break his fall and maintain his fragile hold on reality as a single unbearable thought seared through his mind:

Ghosts are real.

The headset slipped off his ears and dropped to the carpet. His clutching fingers missed the desk and he fell against the chair, toppling it.

Lil called out his name as he sprawled on the rug beside his fallen headset. Her voice sounded small and tinny in his thudding ears. His vision had shrunk to pinpricks. He thought, *I'm dying, I'm dying and I never got to tell her—*

In the Roths's apartment, the words *SECOND PLAYER HAS DISCONNECTED* dripped blood-red down the screen. By the time the zombies had torn her character to shreds, Ben Laramie's heart had disconnected too.

PART 1
TOURISTS OF THE DEAD

Ghostland features hundreds of exhibits and objects haunted by many types of spectral beings (*see* *Index*). Each ghost is made visible using a combination of state-of-the-art Augmented Reality and breakthrough Recurrence Field™ technology. Entire buildings have been disassembled and reconstructed here, alongside the most haunted house in America: Garrote House.

— *Know Your Ghosts:*
A Guide to Ghostland

"Buckle up, babies," Virgil said to his passengers. "You're in for one hell of a ride."

Before any cowards could change their minds, Virgil threw the switch, and the train lurched down into the dark.

— Rex Garrote,
A Roller-Coaster Ride Thru Hell

GHOSTED

Four years later.

BY THE AGE of eighteen, Ben already knew what it must feel like to be a ghost.

In the four years since he and Lil had watched Rex Garrote's house roll down Main Street, he'd spent most of his time either at home or shuttling back and forth to medical appointments. After the open-heart surgery, his parents had pulled him out of school for the coming term. One semester became two, and Ben had fallen behind. His parents had turned to home-schooling.

If he'd been an outsider before—and he definitely had been—he'd become even more of one then. Nobody liked home-schooled kids. That was an undisputed fact. And no one wanted to be reminded of their own mortality, particularly not by someone their own age or younger. So while the town had essentially shunned him, Ben had withdrawn into himself, into books and movies and video games. What little time he'd spent in the outside world, he moved about avoided and unnoticed by other kids, even by many adults.

He'd become invisible. A living ghost, like the main character of Rex Garrote's *Shōki*.

Ben hurried to catch up to Lil outside Duck Falls High when classes let up on Friday. It was the second last week of April and their final school year would soon be over. It was also the day before Ghostland would open to the public. The two of them hadn't spoken in years, although Ben had tried many times, less so this past year. He knew she wouldn't want to see him, but he needed her—his old gaming partner, his former best friend—by his side when he walked through Ghostland's front gates.

Lil kept heading toward the sidewalk, pretending she hadn't seen him. He was used to it. She'd been avoiding him for almost four years now. It still hurt, much more than the others, but he could understand why she wouldn't want him around.

Kids called him names. It embarrassed her to be seen with him.

It was embarrassing to *be* him.

"Hey, wait up!"

Lil flicked a glance over her shoulder and hurried on toward the fenced-in path alongside the school's brick parking lot wall.

"Would you hold up a second?" he called after her.

The two of them alone on the path, Lil finally turned. Shimmering in the afternoon sun, her sleek black hair flipped over the shoulder of her jean jacket. "Leave me alone, Ben." She didn't look up from the cracked pavement between them. "*Please.*"

"I just want to talk, okay?" He felt like an asshole having to chase after her, like some sexual predator or something, when all he wanted to do was talk. "I never get to see anyone anymore," he said. "You know what the worst part of homeschooling is? You can't even watch the clock waiting to go home."

He grinned at that and Lil looked up, locking eyes with him. It looked like she couldn't decide if she should be angry or sad or laugh.

"I have an appointment," she said, her gaze darting away.

"Can I at least walk with you?"

She shook her head. "*Fine.*"

Lil turned and resumed her hurried strut down the path. Some kids Ben didn't recognize ran past the wall at the end of the path. He barely remembered anyone these days, they all looked so different from four years back when he'd still gone to school with them. The three of them stopped there, two burnouts and a jock, and laughed riotously before continuing on their way toward trouble.

Ben wasn't sure if they were laughing at him or Lil or something else entirely. He didn't understand kids much anymore. The few people he talked to these days were adults: relatives, friends of his parents, Grandma Laramie,

Lil's mom, the cashiers at the grocery store.

Lil raised the collar of her jean jacket defensively, hiding behind it.

Because everyone knew The Dead Kid.

Ben Laramie was practically a local celebrity. His heart had stopped beating at the same time Garrote House passed through town on its way to Burt Bucklebee's old farm, land bought up by the late horror writer's estate for a haunting-themed amusement park.

Two other people had died that same day: an old lady had passed away knitting on the porch of the retirement home kids called the "Cripple Castle," and Burt Bucklebee himself, who'd sold his land for a small fortune and had been planning to go to Disneyland, according to the Duck Falls *Squawker*. But old people died all the time. It wasn't every day a fourteen-year-old kid had a heart attack and died only to be revived by a short time later.

A story like that spread fast in a town like Duck Falls.

People called him *zomboy*. *The Dead Kid*. *Freak*.

"I guess you heard about Ghostland?" he asked Lil, trying to sound casual.

She stepped off the path and kept forging onward, crossing diagonally on Maypole Lane. "Duh. Everyone's heard about it. It's all everyone ever talks about everywhere."

"Well, are you gonna go tomorrow?"

She turned on the heels of her black Chuck Taylors. "Why would I want to *go* to that place, Ben?"

Back when the two of them had been best friends, Lil never would have asked him that question. She'd been as obsessed with horror as him. But she was different now. Instead of loose-fitting jeans and oversized T-shirts she wore mostly skirts and dresses. She did her hair up fancy for school instead of tying it back in a ponytail, and she even wore makeup, covering up the freckles across her nose and cheekbones.

He supposed much of that was due to their age. She was no longer the impetuous "tomboy" she'd been when they were fourteen. They were eighteen now, and would be heading off to college soon. Mrs. Roth had mentioned something about a scholarship, though she and Mr. Roth were worried Lil didn't plan to go—that she would rather stay in Duck Falls with her friends than move out and experience the world.

"Why *wouldn't* you want to go to Ghostland?" Ben asked, chasing after her again. "It's like I don't even know who you are anymore, Lil."

"No one calls me Lil."

"*Lilian* then," Ben said, rolling his eyes. "Whatever. I just don't understand what happened with you. Everything was fine until—"

She whirled on the sidewalk, the soles of her shoes grinding the pavement. He could tell what her expression was now: unbridled fury. "I don't want to talk about it, okay? God, it was almost four years ago! Why can't you just *let it go*?"

Ben opened his mouth and closed it again without speaking. "You act like you're the one who died," he said finally, his voice quiet.

"You didn't *die*. Your heart stopped. There's a difference."

"Not according to everyone at your school there isn't," he shot back. "They still call me the Dead Kid, don't they? They still say I'm a freak?"

Lil gaped at him a moment, seemingly undecided whether to tell him the truth and break his heart or lie and make it worse. Instead, she turned and stormed away.

Ben stood in the middle of the road watching her walk away from him until a car honked. He turned to the woman in the minivan, scowled at her and her baby in the backseat, then crossed to the sidewalk.

Lil had headed left toward her therapist's office. Ben turned right toward home.

LILIAN SAT in the waiting room at Dr. Wexler's office, trying to distract herself from thinking about her encounter with Ben.

She'd done such a good job avoiding him during the past few months that it had almost felt like he didn't exist, like he had actually died that day, like his mother hadn't come home early from showing a house on the other side of town just a few minutes after Ben's heart stopped beating and found him lying on the bedroom floor in a mess of horror books, his hands clenched into fists and his face blue.

Not having to worry about running into Ben around town, kids at school or her parents asking about him—as cold and morbid as that sounded, Lilian couldn't help feeling comforted by the idea.

She looked around the waiting room. A middle-aged white woman wearing a sari sat opposite her, texting, the phone clicking with each keystroke. She recognized the woman as a local yoga instructor, a recent addition to Duck Falls from somewhere out of state.

The TV in the high corner showed the Baltimore traffic report. Behind the woman in the sari was a poster that said *IT'S ALL IN HOW YOU LOOK AT THINGS*, spelled out like an optometrist's eye chart. This hokey motto never seemed to jibe with what Lilian knew of Dr. Wexler, a no-nonsense woman with a take-charge-of-your-own-destiny attitude. It felt incredibly phony, which Lilian—much like Holden Caulfield—couldn't stand.

Her gaze drifted to the magazines. On top of the stack, the *Time* magazine cover showed a somber-looking black woman with large, thick-framed glasses. Lilian knew her from television. The most-recognizable woman in the Western world at the moment wasn't a reality-TV star or royalty or even a politician. The title beside the photo said *LIFE AFTER DEATH? The Brilliant Mind Behind Ghostland's Terrifying AR Experience.* She was Sara Jane Amblin, inventor.

Lilian reached for the magazine to flip it over but the woman in the sari got to it first.

"Oh, did you want to read it?" the woman asked. "I just *love* her."

Lilian shook her head brusquely. The woman gave her a befuddled look and opened the magazine, so that now the inventor of the "Recurrence Field" was practically staring at Lilian across the table.

I guess you heard about Ghostland.

Everything was about Ghostland these days, and when people spoke of Ghostland what they meant was ghosts. Death. The paper-thin veil between the living and the dead. Lilian couldn't go a day without seeing something about it on the news or hearing someone talk about it in class. Last month the school board had voted on whether to add a special "death and bereavement" segment to the health curriculum. Despite a few protests from religious types it had passed unanimously, which meant next year kids would be hearing about it in gym class too.

Death was all Dr. Wexler wanted to talk about. And it was all Lilian could think about whenever she saw Ben.

On the television in the high corner of the room, an ad came on for a *60 Minutes* interview with the inventor. Lilian had to clench her fists to stop from screaming in frustration.

The door to the inner office opened then, drawing her attention away from the TV. A meek-looking man with an undercut and an ugly wolf-howling-at-the-moon sweater stepped out, followed by Dr. Wexler herself.

"See you next week," the therapist said.

The little man nodded with his chin tucked into his shoulder and shuffled out of the office.

"Lilian?"

Lilian stood up, gathering her knapsack and phone. Dr. Wexler smiled and ushered her in. She was fortyish, maybe as old as forty-five. She kept her dark hair in a high bun and wore awful primary-colored pantsuits. But she always had her nails painted black, which Lilian thought was an odd juxtaposition, considering her age and the I'm With Her outfits. She was also undeniably attractive, Lilian thought, for a woman her age. The fine lines on her pale skin accentuated her good looks rather than detracted from them. In a little black dress, she might even look elegant.

Lilian stepped into the office and flopped down on the sofa. Dr. Wexler closed the door and slipped into the hard, boxy accent chair she seemed to prefer to the plush leather rolling chair behind the desk. Lilian thought she liked it because it was as severe as Dr. Wexler herself.

The therapist crossed her legs at the knee and laid her hands one on top of the other upon them. "So... how are you feeling today?"

"Fine," Lilian lied.

"Fine? Not good?"

"Your powers of deduction are astounding, Doctor."

The therapist engaged her in a staring contest. Lilian blinked first.

"I saw Ben today," she said.

"Ah," the woman replied.

"*Ah*?"

"Ah." Dr. Wexler nodded. "As in, *I see*."

"You don't have to be sarcastic."

"What should I be then, Lilian? You've been coming to me for six months. I ask you questions, you shy away from them. Your parents are concerned about you. So am I."

"Why? My grades are good. I'm not a loner or a bully or bulimic or cutting myself. I haven't shot up my school. Why can't you all just let me live my life in peace?"

"Because you refuse to live," the doctor said. "Because you're so terrified of death it's crippled you on an emotional level. Lilian, you refused to sit *shiva* when your grandmother passed last March. Just two weeks ago, you had a panic attack after seeing a dead baby bird on the sidewalk. I can *help* you. But only if you let me."

Lilian made to speak but couldn't find the words.

After a moment of silence, Dr. Wexler asked, "Is that new?"

Lilian wasn't sure what the woman meant until she realized she'd been fiddling with her bracelet again. Her parents had bought it for her when her Mid-Year Report had gotten her a scholarship. It wasn't much—though it had cost more than they should have been spending—but she'd come to fidget with the beads in times of stress like a Catholic counting her rosaries.

She gave the doctor a brief synopsis of its origin, adding, "I think it's kind of ugly but I don't have the heart to tell them."

"Really. I think it's quite lovely."

"You want it?" Lilian held out her wrist.

"I can't take your bracelet, Lilian. But I wonder..."

Now it was Lilian's turn to raise her eyebrows. "Wonder what?"

"I wonder if the reason you don't like it is its intended purpose, to encourage you to accept this scholarship. And I wonder if your fidgeting with it in times of stress might be tied to your hesitation to make difficult choices about your future. Stanford is all the way across the country, after all. Far from the security and stability of home."

"Gee, you really got me figured out, Dr. Wexler."

The therapist grinned. "Do you know the five stages of grief?" she asked.

"*Ben never died*," Lilian told her for the hundredth time.

Dr. Wexler gave her a thin smile. "Not in a literal sense. But to you, he *was* dead. For those few minutes. You heard his mother screaming. Giving him CPR. And you couldn't do anything to help him. It was a devastating emotional trauma, Lilian. People experience similar traumas all the time and never bounce back from them." The therapist tented her black fingernails against her lips. "But because Ben *didn't* die, because his mother was able to *revive* him, you've never allowed yourself to experience the final stages of your grief. You're caught in a limbo of denial."

The therapist arched her left eyebrow. In Dr. Wexlerland, this was an indicator of amusement or self-satisfaction, depending on the context. Here it meant she was pleased with a turn of phrase. "In a sense, it's as if a part of yourself died that day," she finished.

"That's what Ben said," Lilian muttered.

"What else did he say?"

"It's stupid."

"Well, it's obviously disturbed you. Last week you weren't so evasive."

"He wants me to go to Ghostland," Lilian said. "See? Stupid."

"Hmm."

"What do you mean, *hmm*?"

"I don't think that's stupid at all. In fact, I think... it *might* be helpful."

"Helpful?"

Dr. Wexler held up a hand. "Hear me out. From everything I've heard about Ghostland, it appears to be a safe environment where you could—" She smacked her tongue against her lips, choosing the right word. "—*explore* your phobia without harm," she finished. "Exposure therapy, let's call it. Of course, I'd be happy to escort you myself—"

"*Escort* me? My parents can barely afford to send me to you once a week."

"I'm not asking for money, Lilian." The therapist uncrossed her legs and rested her elbows on her knees, positioning herself closer to the sofa where Lilian sat, and eyeing her with intent. "Consider it a field trip. As a matter of fact, I've been thinking I should go there myself out of professional curiosity anyhow."

"Great." Lilian rolled her eyes. "Then maybe you could pay *me* to go."

"So it's settled? Shall I pencil you in for nine A.M., tomorrow?"

Lilian shrugged, shaking her head. "Why not? What's the worst that could happen, right?"

HOME

ON SATURDAY MORNING, Ben woke early and tucked some things he'd need for the day into his backpack. He wasn't sure if security at Ghostland would allow him to bring in outside food but it was always best to be prepared. Few things were worse than getting stuck somewhere without anything to eat. As his mother would often remind him, he was a growing boy. Although it was obvious to anyone other than his mother that he hadn't grown an inch since his heart attack.

He packed smartly: a bottle of water, a few Nature Valley granola bars, his heart pills (which he hoped wouldn't be necessary, but it was better to be safe than sorry, particularly when "sorry" potentially meant death), some carefully folded tissues, a pack of gum and a small bottle of hand sanitizer. He'd tucked an ex-library copy of Rex Garrote's *The House Feeds* in with them, along with a few other items he'd need, zipped discretely into the false bottom of the bag.

In the past few weeks, he'd spent a fair amount of time researching Ghostland's exhibits. In spite of its reputation, the house he and Lil had watched roll down Main Street nearly four years ago, Garrote House, wasn't its most haunted exhibit. Bright Falls Sanitarium was worse, allegedly haunted by the spirits of nuns and mental patients and homicidal doctors, and Fontaine County Correctional, which had put at least a hundred men to death in its day, was said to be the most haunted place in America. The Ghostland website claimed you could take part in an "authentic prison riot experience," except that all the prisoners were long dead.

He'd also spent a good portion of the past year re-reading Rex Garrote's novels. It had been a long time since he'd read some of them, and he'd forgotten how brutal they were. Garrote's prose was a blunt-force strike directly to the fear center, mainly concerning itself with flawed protagonists, sex and death. Everything else was set dressing.

The books were also unrelentingly bleak. Where writers like King and Koontz often had upbeat endings, Garrote seemed to prefer grim. The few victories his characters found were only ever no-win situations, the casualties so high and the toll on the main characters so devastating that even winning felt like a loss. Out of Garrote's twenty-six novels, only three had what could properly be called a "happy" ending. Of those, the outcome of two were ambiguous at best.

Even most of the old television episodes ended with a decapitation or a vicious mutilation or a *Hamlet* "everyone dies at the end" scene. Ben suspected this was why *Rex Garrote's Ghost World* had only lasted a single season. People preferred upbeat endings, where everything was tied up with a nice little bow, at least before the advent of binge-watch TV.

Garrote despised loose ends, though bows had been much too pretty for his liking. Instead he'd snipped them with blood-drenched garden shears.

"Garrote seems more interested in scenes of gratuitous exploitation than such tedious trappings as character development and believable plotlines," a *New Yorker* reviewer had suggested in the late-'90s. This was exactly what Ben had enjoyed about the books. No wasting time building characters or setting the mood, just rocketing straight into blood, guts and sex, gritty characters and grittier settings, with a liberal sprinkling of '70s "jive talk" that often didn't fit the characters. If Ben's parents had gotten any inkling of the kind of stuff he'd been reading, they would have lost their minds.

During his lifetime, Garrote had been called both "the scariest man in America" and "profoundly disturbed and disturbing" by critics, and following his death it had turned out they were right. In the summer of 1999—the beginning of a renaissance period for horror movies, the year of *The Sixth Sense* and *The Blair Witch Project*—the writer's remains had been found cremated within his private library at Garrote House. There had been no sign of a break in, according to the detectives in charge of the investigation. Either the killer had known him and been given access to the house, or the writer had given new meaning to "profoundly disturbed" by burning himself alive.

Self-immolation, Ben thought as he double-checked the zippered bottom compartment in his backpack. *That's about as bleak an ending as you can get.*

He threw the backpack over his shoulder, steeling himself, ready for what came next yet secretly hoping it wouldn't come to that.

From now on, he would need to be alert. His mom was in the backyard hanging up the wash, and his parents had no idea about his plans. Normally she would be preparing for an open house in the afternoon, but the market had taken a bit of a hit after the Ghostland upsurge. There were no new listings for her to show this weekend and so she was home, which meant Ben would have to sneak out of the house.

He was under strict rules. No strenuous activities. No excitement. No life.

And horror... well, his fixation with horror was, to his parents, *disturbed.*

If they thought he was disturbed now, he could hardly imagine how they'd feel after today.

Following his sudden cardiac arrest, he'd narrowly prevented them from burning all of his horror stuff on the back lawn by faking a fainting spell. His mom had taken pity on him, figuring that the stress getting rid of it would cause him wasn't worth the trouble.

What they didn't understand was that horror *comforted* him. Ben had once attempted to compare it to driving by a car crash. "Okay yeah," he'd told his mom and dad, "you might see a few mangled bodies, and there's probably gonna be a big explosion at the end, but you're driving by with the doors locked, safe inside."

His parents hadn't liked the analogy. They also hadn't appreciated the thought of him driving.

As he creeped out of his bedroom, he heard his mother coming back inside the house. The bottom stair creaked and his heartbeat felt like thick liquid. He tiptoed toward the front door, twisted the knob ever so slowly and stepped out into the sunshine.

"Ben?" his mother called from the kitchen. "I thought I asked you to load the dishwasher?"

Ben closed the door carefully and hurried up the walk. Dr. Wexler's office was less than a mile across town. He'd have to take side streets to avoid bumping into one of his mother's many friends and acquaintances, but with any luck he'd be well inside Ghostland before she even realized he was gone.

He just hoped he could make it through security with what he had tucked inside his bag.

LILIAN'S DAD shuffled into the kitchen while she was scraping butter onto a charred piece of toast. She brought it to the small kitchenette table and sat down in front of her mug of black coffee.

"You're gonna get an ulcer with that breakfast," he said.

Lilian hummed in reply and blew on the mug's hot contents.

Hiram Roth sat down beside her, resting his bristly chin on his hands. Without looking up from her coffee, she sensed her father's stare. She knew he would just keep staring at her until she acknowledged him, even if it meant he'd be late for a job.

"Yes, Daddy?" she said in her best I'm-an-innocent-princess voice.

His lips spread in a goofy smile. "My angel," he said. "Look at you. Grown up so fast. You got your mother's good looks, thank God. Can you imagine that gorgeous face with these eyebrows?"

Lilian sipped her coffee, trying not to grin.

"You know, when you were just a little one, Mom and I always told you you'd go on to do great things. We hypnotically suggested it in the crib, I kid you not. *Baby Einstein* is an amateur compared to your mother and me."

She couldn't help but chuckle behind her hand.

"And now here you are, just a few months from graduating. Heading off to college. Oh, to be young again," he said dramatically, channeling his brief high school theater experience. "We're so proud of you, honey. But you know, you don't have to go to UMD just because it's close. Don't be afraid to spread your wings a little."

"I want to stay here with you and Mom."

He shrugged up his shoulders, giving himself a bushy double chin. "But there's maybe better opportunities for you elsewhere. What about that scholarship? Your *zayde* went to Stanford, you know."

"We can't afford Stanford, Daddy."

"The scholarship will cover a good chunk. You could always take out a student loan. Mom and I will cover what we can. We just don't want you to limit yourself, sweetie. Duck Falls is a wonderful little town, but this place

—" He made a grabbing gesture with his blistered working man's hands. "—it sucks you in. It's like a black hole. Your mother and I, when we schlepped out here from Baltimore, it was an exciting opportunity for the both of us. A place to lay down roots. When the business folded, we had nothing. *Less* than nothing. We did the best we could. We provided for you so that you could have more than us, sweetie. Not the same. Not less."

"Dad..."

"Don't you *Dad* me, young lady. I'm not trying to guilt you. I'm just imparting some fatherly words of wisdom."

He took her hands in his, squeezing them tenderly in his rough palms. His gaze fell on the bracelet they'd gotten for her (*Let's be real, Dad got it for me, not Mom*, she thought), and he smiled.

"Don't get your butt stuck in Duck Falls, honey. Mom and I don't regret coming here, not for a day. We could have gone back to Baltimore with our heads hung low when the economy went down the crapper. But we believed, we *still* believe, raising you here—not pulling you out of school, away from your friends, away from your life—that it was the right thing to do."

He squeezed her hands once more and let them go. She returned his smile, although the last thing she felt like doing was smiling. Her father had spent the last ten years working odd jobs around town while her mother Maddy busted her hump for tips at the 86 Diner, all so Lilian would never have to experience the minimal discomfort of adapting to new surroundings.

She felt awful about it, always had. The coffee churned in her empty stomach. Afraid she might vomit if she didn't eat something fast, she wolfed down her toast.

"You've got my appetite, though," her dad said with a grin, rising from the table and lifting his gut to fasten the tool belt over his hips. "Okay," he sighed. "Have fun at the spooky park today, sweetie."

"Doubt it," she muttered.

"Well, if you can't have fun at least stay safe." He cupped the back of her head and kissed her on the brow. "Love you, sweetie."

"Love you, Dad," she said, watching him leave.

As he twiddled his fingers at her through the apartment doorway, a sudden inexplicable thought struck her, causing her hand to quiver and slop the contents of the mug.

What if this is the last time I see him?

No reason at all to think he would die on the job, nor that something might happen to her at the "spooky park." It was just an awful, unshakeable thought, one of many "intrusive morbid thoughts," as Dr. Wexler called them, that she'd experienced ever since Ben had almost died.

But this one... it felt prophetic somehow. Like she should heed its warning.

You're just scaring yourself, she thought. *Trying to convince yourself not to go today. Just suck it up, Dorothy. In a few hours you'll be home again.*

Determined to push through her anxiety and confront her fears, if only to prove Dr. Wexler wrong, Lilian got up from the table, poured the rest of her coffee down the sink, and headed to her room to get ready.

#GHOSTSRPEOPLE2

LIL AND HER therapist were waiting outside the medical center at the western edge of downtown when Ben arrived, and he was already sweating through the underarms of his T-shirt and where his backpack rested. Spring had come on fast and strong in Duck Falls, with summer already close on its heels. He hoped it wouldn't be too hot this year, otherwise he'd have to spend most of the summer in the air-conditioned house. Extreme heat could be harmful for the heart of a normal person. For someone like Ben—*A freak*, he reminded himself—it could be deadly.

He hated not being able to do the things "normal" kids did. Just once he would have liked to go cliff diving at The Hole during the summer like other kids his age, to play an organized sport, or have a few beers at a bush party, make out with a girl and get into a stupid drunken accident on a four-wheeler. Normal kids took these things for granted.

Everyone except for Lil.

She could do whatever she wanted and actually *chose* to do nothing. It was one of the main reasons her parents had put her into therapy. On Fridays Ben's mom would let him go to the 86 Diner for lunch, just to get out of the house. He'd spend an hour there, eating a BLT and fries ("Something fatty once a week," his mom often cautioned), and reading a book until Mrs. Roth took her break, at which point she would dispense the latest gossip about her daughter. After lunch he would pack up the second half of the sandwich and take it to Marshland park where he'd watch the ducks, and toss sticks loaded with pine sap into the creek to watch them motor off toward the Potomac. Toward freedom. Somewhere far, far away from Duck Falls that Ben himself would likely never visit.

Because of his chats with Mrs. Roth—she let him call her Maddy—Ben knew why Lil was seeing Dr. Wexler. He also knew she hadn't agreed to come today because she wanted to catch up with her former best friend. He could deal with that. He could even tolerate her therapist tagging along as chaperone. Ever since he'd heard about Ghostland, he'd wanted—*needed*—Lil to come with him on opening day. He wouldn't have the courage to do what he needed to do without her. Whether she'd been pressured to come by her parents and therapist or not, she was here and they were going together. It was all that mattered.

Rex Garrote, dead almost twenty years, was still in that house.

Watching.

Waiting.

What exactly was he waiting for? Ben had no idea. What he did know was that none of the dozens of reports of hauntings in Garrote House had ever mentioned the presence of the man himself. Of all the ghosts visitors had claimed to have seen in that house, Rex Garrote had never been among them.

Except he was there. Because Ben had seen him.

He needed to prove that to himself as much as to Lil.

And then he would need her help, to find the dark heart of that house and coax its master out of hiding. He needed his teammate, his partner, this one final time, to get rid of Garrote once and for all.

DR. WEXLER STEPPED out of the driver's side as Ben approached the car. He recognized the woman from around town but judging by her profession and her typical attire of primary-colored conservative pantsuits, he'd expected her to drive an expensive foreign sedan. Instead, she drove a blueberry-blue Honda hatchback and wore tight fitting blue jeans, white dock shoes and a loose flower-printed shirt. Her hair, normally in a tight bun, lay in shiny auburn waves over her shoulders.

Lil remained seated in the back, turned away from him, just the back of her head visible through the rear window.

"You must be Benjamin," Dr. Wexler said, extending her long, slender fingers, the nails painted black. "I'm Allison."

"Ben," he said, shaking her hand. He wasn't sure if he was overheated from the walk or if the therapist just had extremely cold hands. Either way he was glad when she released him from her grip.

"Sit wherever you like," she said.

"Can I drive?"

Dr. Wexler—Allison—smirked. "Anywhere except the driver's seat," she clarified.

His shoulders slumped. He hadn't expected her to say yes, but she had inadvertently made the offer. He went around the back to open the front passenger door and flopped into the seat, tossing his backpack between his feet. The lighter fluid sloshed with a hollow metallic glug that he hoped no one had heard aside from him.

Dr. Wexler—*Allison*, he reminded himself again—sat behind the steering wheel and buckled in. Ben smiled over the back of the seat at Lil. "Hey, Lil. *Ian*. Lilian," he said, wincing at his awful attempt to cover.

Lilian gave him a tight smile and rested her chin on a fist to peer out her window.

"So, who's excited?" he asked. "I know I am."

He caught Lil's eye roll in the side-view mirror and chose to ignore it.

Allison dropped the car into reverse and backed out into the empty road. "Actually, I am a little excited. Academically speaking," she added with a cautious glance in the rearview, as if she might be worried Lil disapproved of her public display of enthusiasm. "How are you feeling, Lilian?"

"Do we *have* to do this?" Lilian asked.

"If you've changed your mind—"

"I'm going, okay? Jesus! I just don't want to pour over every goddamn twitch in front of... *him*." She stuck her chin out antagonistically toward the back of his seat. "Can we just pretend you're not my therapist? For today?"

Ben saw Allison purse her lips before she nodded, controlling her anger. "Okay," she said. "Today I'm just Allison. We're just three friends going to an amusement park together."

"Great," Lil muttered. "Let's just get this shit over with."

Allison arched an eyebrow at Ben. He hid his grin from Lil.

About half a mile outside of town, past the old fairgrounds and the abandoned sawmill, traffic from the highway exit slowed their progress to a crawl. Allison tapped her glossy black fingernails impatiently on the steering wheel. After a few minutes, the heat inside the car started to make him nauseous. He zipped down his window, hoping he wouldn't need to puke.

He spotted several people walking in the ditch along the line of cars just up the road, headed toward the car. It reminded him of the times they'd gone into the city and panhandlers had approached, asking for change. A young woman with orange dreadlocks and baggy camo cargo shorts that hung down below her knees knocked on the window of the Suburban ahead of Allison's car. The woman held a stack of pamphlets in her left hand.

The passengers of the Suburban ignored her. She muttered something snarky Ben didn't catch, marched up to his window and held out the pamphlets.

"Ghosts are people too," the protestor said in monotone. "Help put an end to the slavery of innocent non-corporeal beings."

Ben reached for the pamphlet. The image on the front was a cheap rip-off of the *Ghostbusters* logo. Printed along the slash was the hashtag #GRP2 (which Ben guessed stood for "Ghosts are People Too"), along with the slogan "Free the Ghosts."

"My dad says that's communist propaganda," Lil piped up behind him.

"It's not *propaganda*," the woman snapped, her light blonde eyebrows with multiple piercings knitting together. "How would you like it if they put your grannie on display in there?"

"I'd be proud of her," Lil shot back. "Nana always wanted to be in the movies."

Ben heard the activist's teeth clack together in stunned silence. He took the pamphlet, thanked her and zipped the window back up before snorting laughter.

Lil shook her head. "You're just encouraging them."

"I dunno, I kind of like this stuff." He folded the pamphlet and tucked it into one of the pockets of his backpack. "I've got a whole bunch of those Chick cartoons at home."

Lil lurched forward in her seat, her interest suddenly piqued and her seatbelt cranking. "You collect *hentai*?"

"No!" he responded too quickly, his cheeks already burning in embarrassment. "They're religious tracts. Jack Chick was the guy who made them. You've probably seen them online, those funny comics about how everybody's going to Hell and stuff."

"Except now we're all aware there is no such place," Allison said in a sagely tone.

"Nobody's proven that yet." Ben pointed up ahead. "They sure don't seem to think so."

Lil and Allison looked where he was pointing to a mass of protestors in the high grass alongside a farm fence, angrily waving signs. Judging by the differences in their attire, opposition to Ghostland seemed to be one thing Christians, Muslims, hippies and Mormons all could agree on. It reminded Ben of the time he'd been forced to cross a picket line at the hospital to get to his appointment when the nurses had all gone on strike.

"Oh boy," Allison muttered.

The few signs Ben could read from the distance bore bold statements in big bold letters: DEATH OBSESSION IS DEMONIC POSSESSION, LIFE IS BEAUTIFUL - GOD IS LIFE, and SAY NO TO DEATH-ED CURRICULUM! Everyone appeared to be chanting something, but he couldn't make out the words until the traffic allowed the car to roll forward a few more yards and it came to him on the light pre-summer breeze.

"'Afterlife is not your choice,'" he chanted along with them. "'Suicide is not my voice.'" He laughed again and shook his head.

"Bunch of whackos," Lil muttered.

"It's important to be respectful," Allison said, peering back over her shoulder. "Humanity is undergoing a massive shift in consciousness right now. Some people aren't capable of handling it as well as others."

"Like Twitter," Ben said, and Lil laughed.

"We just have to allow each other space to deal with it in our own way," the therapist continued, ignoring his quip.

"But they're stuck in denial," Lil said. "Ghosts are real. Heaven is a fairy tale. Everyone just needs to accept that and move the fuck on."

"Don't forget Midian is where the monsters live," Ben added.

Allison arched an eyebrow at him again, not getting the joke. He knew Lilian got the reference—they'd seen the movie together—but she ignored him.

After the turnoff to Ghostland, traffic separated into four lanes onto property that had once been Burt Bucklebee's wheat farm. Golden fields of remaining straw swished and swayed on either side of the road in the warm, sweet-smelling breeze.

Ben slipped half out the window to get a good look at Ghostland up ahead. He'd seen them raising the massive gray wall surrounding its perimeter on drives to the Wal-Mart in Hagerstown and doctors' appointments in Frederick. At the time, no one in Duck Falls had any clue what they were building out here, even most of the people constructing it. All they'd known was that the land had been zoned for a theme park, and that the trucks passing through town seemed to be hauling a lot of old buildings and vehicles and all kinds of junk out there. Nobody had any idea the park would change the shape of the world forever.

But with Ben's overactive imagination, he'd always believed something wasn't right about it. Driving by at night on the way back from shopping or a movie, he'd taken to holding his breath like a superstitious kid passing a cemetery. He supposed he might have somehow sensed Garrote House stood behind the wall even before the announcement was made... and that Rex Garrote's ghost had been biding his time within its haunted halls.

Peaked roofs and gables were visible now over the top of the wall, along with cable car support towers and what looked like a glass-roofed rotunda, which Ben thought must belong to the prison. With a bone-white moon casting shadows over these lonesome fields, it might have made him anxious. Approaching it in daylight was no more fearsome than a trip to the zoo.

Still, the wall itself troubled him. Why would they build it so tall you could hardly see the exhibits from the outside? Wasn't that part of the joy, the excitement of driving to a place like this, the anticipation that seeing all of the exhibits and rides in the distance brought? The flat cement wall seemed like something they would put around a government agency, not a theme park.

"Is that wall to keep people from sneaking in, you think?" He glanced over his shoulder at Lil. "Or to keep the ghosts from getting out?"

Dr. Wexler arched her eyebrow at him.

"Just curious," he said.

"They're gonna *gitcha*, Ben!" Lil shouted in his ear, grabbing his shoulders and shaking him briefly. He jumped in his seat and almost bumped his head on the ceiling, while Lil laughed like it was the funniest thing ever.

"Ha ha," he said. "You're *so* funny."

A bored woman dressed in an orange safety vest waved them toward the right. The parking lots to their left were already full. Allison turned into the mostly empty lot and quickly found a space. She unlocked the doors and the three of them got out of the car to stretch.

Ben's backpack glugged again as he slung it over his shoulder. Allison said, "I don't think they'll let you in with that."

For a moment he felt caught, thinking she must have heard the lighter fluid slosh. Then he relaxed. Even if she had heard it, she would have mistaken it for a water bottle.

"I have to bring it," he said. "It's got my pills and whatnot. I have a medical alert bracelet." He raised his arm to show her. The little metal medallion tinkled on its chain. Sunlight flashed off its surface into her eyes, and she squinted. "Sorry," he said, self-consciously tucking his hands into the pockets of his shorts.

"Well, it can't hurt to try," she said. "Worst that can happen is we have to leave it at the security desk, right?"

"Right," Ben said, thinking, *Please, please, don't let that happen.*

Lil said, "Less talky, more walky," and started toward the entrance.

Allison arched an eyebrow again before taking up behind her. Ben grinned and followed them, excitement hurrying his step.

Whatever lay ahead of them, he suspected the park itself would be a riot.

WELCOME TO GHOSTLAND

LILIAN GROANED AT the endless lines streaming toward the entrance. She hated crowds. They were part of the reason she disliked going to places like this. It wasn't that she was agoraphobic, she just wasn't very comfortable around strangers. Strangers were unpredictable. They were racists and homophobes and rapists. They were random acts of violence just waiting to happen, at least according to the news.

An old man in a vintage sport coat, his balding head covered by a crumpled tan fedora, glanced over his shoulder as Ben and Dr. Wexler stepped into line behind him. He gave Lilian a brief smile. Not creepy, just friendly. At eighteen, she'd already dealt with enough creeps to know the difference.

"I hope they have maps," Ben said, out of nowhere. "I like maps."

"You *like* them?" Lilian said.

"Yeah. I collect them," he said, beaming at her. He still had bedhead and a big hunk of hair on his crown bounced when he nodded.

Lilian couldn't help but snort. She'd almost forgotten how tragic he was. When they were kids it hadn't bothered her, she'd found it kind of endearing. Almost adorable. As a new adult Lilian had tried her best to block that period of her life out of her memory. She didn't need to be reminded of how much of a dork she'd once been herself. "*Cool*," she lied.

Ben's smile widened. He had no idea she was being sarcastic.

"You know, I collect maps myself," the old man in the crumpled fedora said. "Ptolemy, Kwon Kun, Mercator—all the big names. I'm a cartography buff, I guess you could say."

"Cool," Ben said.

The man grinned, flashing yellowed teeth too straight not to be dentures. "I don't know if it's cool but I know what I like."

"It's definitely *not* cool," Lilian said. The old man gave her a dismissive look before returning his attention to Ben.

"Those ancient sea maps are the best," Ben said. "The ones with all the sea creatures and stuff. *Here be dragons*."

The man smiled. "The Lenox Globe. That one's my favorite too." He reached a calloused hand past Lilian. "I'm Stan. Stan Beadle."

Ben shook it. "I'm Ben. This is Lil."

She gave him a look of death.

"*Lilian*," he backtracked quickly. "And that's Dr. Wex—I mean, Allison."

Stan shook Dr. Wexler's hand, repeating his name with a "Pleased to meet ya," then held his hand out to Lilian. She shook it, mumbling *nice to meet you* even though she didn't need anyone encouraging Ben's dorky behavior, and worse, the man's palm was clammy and gross. It took all of her will power not to squirt Purell into her hand from the travel bottle she carried the second he let go.

"So what, may I ask, brings the three of you to this most inauspicious of places?"

Ben said, "Well, ever since I first heard about it I've wanted go, and Lilian's therapist told her she needs to—"

Mortified, Lilian smacked him on the shoulder before he could reveal anymore of her secrets to some random old dude.

"Hey! What was that for?"

"For being the world's most clueless dumbass."

"Hey, kids, I don't want to get into the middle of a thing," the old man said. "Just curious is all."

"The kids—" Dr. Wexler caught the burning hatred Lilian fired at her and paused to course correct. "*Lilian and Ben* were going, and I agreed to chaperone. I intend to write a paper about the possible effects of repeated Recurrence Field usage on the human psyche."

"Wow, that is..." The old man nodded, frowning in obvious confusion. "That's a lot of big words is what that is."

Dr. Wexler shrugged. "It's not really for laypeople."

"I wouldn't consider myself a layman, exactly," the old man said. "I'm a detective." He shrugged. "Used to be. Retired six years come September. And I got nothing against shrinks personally, I just think you folks spend too much time tinkering with what's in here—" He tapped the wild gray hairs at his temple. "—that you sometimes miss what's right in front of your eyes."

Lilian watched her therapist's cool demeanor crumble like a day-old babka and she began to smile, enjoying this new dynamic. After all the time she'd spent trying to crack Allison and this old fart had figured her out in less than a minute.

"Agree to disagree," Dr. Wexler said, her jaw clenched.

"Fair enough." The line moved up and the detective shuffled forward. "Anyways, I'll let you get back to it. Pleasure to meet you all." He drew a thumb and forefinger around the brim of his cap the way cowboys did in old movies and turned to face the front.

Dr. Wexler continued to stare at the back of the man's head for a moment, her lips pursed. Lilian had never seen her therapist angry before, no matter how many times she'd provoked the woman herself. The detective had thrown some serious shade.

"*Respect*," she muttered to herself. Ben gave her a strange look. "What are you looking at, map boy?"

BEN LAUGHED off Lil's attempt to embarrass him. The comment contained a layer of irony he wasn't sure she had noticed herself: when they used to game together she had constantly checked back with the map. If anyone should be called map-anything it was her.

He also realized Lil didn't like her therapist much. The detective had clearly upset Allison and Lil had seemed to take pleasure in it. He just hoped the two of them would loosen up a bit once they got into the park. Bad enough he had to deal with Lil's bitingly sarcastic animosity toward him without getting in the middle of a blood feud between doctor and patient.

Ten ticket booths stood out front of the large gateway with a separate line for each, although as they got closer together it almost seemed like one big mob. Their line progressed and Stan, the detective, bought his ticket and stepped into the park, tipping his hat in their direction.

Ben liked the man and wondered what he might be doing here. Some futurists had predicted a portable version of this new technology—the "Recurrence Field"—could be used to solve murders someday. Maybe Stan Beadle had considered this, and was here to crack an unsolved case. Ben thought that would be cool and it made him like the man even more.

Allison asked for three adult tickets. Lil gave her an appreciative look, as if she expected her therapist would try to get them in as "under 12." A ginger-haired kid Ben recognized as a former schoolmate from a grade ahead slipped three pairs of glasses through the brand-new Plexiglas window. Each headset was fitted with a pair of earbud headphones, and had two buttons on the right arm: a green Y and a red X. A small, blue LED power indicator glowed beside them.

Ben had only ever done virtual reality once, in a demonstration at the mall. The images had been blurry, the headset too tight and hot, and the machine had jostled him so much he'd jumped off and rushed to the bathroom, feeling like he was going to throw up. These headsets were supposedly AR—augmented reality—and not VR. The main difference was that AR overlaid images and sounds over reality itself, whereas virtual reality was entirely computer generated, essentially a 3-dimensional video game. With VR, the headsets were often large and clunky. Most AR games were only playable on mobile devices, like phones and tablets. According to an article Ben had read in *Wired,* these headsets—not much bigger than those blue-blocker sunglasses they sold to old people during daytime court shows —were brand new tech only available at Ghostland.

He glanced over at the booth to his left, where a group of Asian goth kids about his age slipped on their headsets, speaking excitedly to each other. With an I'm-about-to-lose-my-patience smile, the matronly ticket-seller at their booth asked, "Now ya'll wanted your language set to Japanese, is that right?"

Multiple languages? Ben thought. *They really went all out with this.*

Allison examined her pair. "How do these things work?"

"Those are your AR glasses," the kid in their booth said in a vocal fry drone. "You'll only be able to see and hear the exhibits when you wear them. If you start to feel disoriented or sick to your stomach, just remove the glasses. But you should get used to them pretty fast. Now you're gonna see some blue text on the inside of the glasses sometimes. That's the Heads-Up

Display. To interact with it, just press the green or red buttons on the side of the glasses."

"Do we get a map?" Ben asked.

Lil chuckled and shook her head.

The kid pointed behind himself. "Maps are available just inside the gate there."

Ben watched Lil slip the glasses over her ears and pop in the headphones. She looked around, blinking rapidly. Her eyebrows rose above the black rims. "How do I look?" she asked.

He swallowed a hard lump, unable to stop the blush from rising in his face. "They're you," he said, and busied himself fiddling with his own pair. Ben hated his goddamned cheeks. Once they started burning, trying to stop it just seemed to make it worse. He glanced up at Lil, hoping she hadn't noticed, but she was looking at him with a sly grin as if she knew exactly what he was thinking.

"Be careful," the kid in the booth said. "When you first put 'em on it might be a little scary." He blinked at Ben. "Hey, wait... aren't you the Dead Kid?"

"Boo," Ben said.

As he passed through the gateway, he noticed a metallic inner door he supposed could be raised and lowered like a castle gate. Heavy security. On the other side of the wall, several security guards stood patting down guests as they entered the park. The guard at the head of their line looked Hawaiian, with a tribal tattoo on his neck poking out above his collar. His name badge said NIKO.

Ben stepped up to the gate, trying not to look suspicious. Lil and Allison were already waiting for him just beyond the checkpoint. If he got caught now, they would have to go on without him.

"I need to check your bag, sir."

He heard a slight slosh of liquid as he handed over his backpack. The guard eyed him a moment and began rooting through it. He pulled out the water bottle first and shook it, causing a similar sound. He nodded and returned it to the bag, which caused the pill bottle to rattle. Niko scowled and pulled them out.

"What are these?"

"I have a heart condition," Ben said sullenly.

The guard nodded and tucked the bottle back inside. "Technically, we're not supposed to let in people with heart conditions."

Ben felt his chest tighten, and if he hadn't been in public he might have started crying. He was so close. If they didn't let him in after all the time he'd spent preparing, after getting Lil to come with him despite not wanting to, he didn't know what he would do.

"That's against my constitutional rights!" a tubby man with a bushy beard and a NO FEAR T-shirt shouted from the next line over. He wore an old Orioles cap over his shaggy brown mullet. "I've got my concealed carry permit right here, man!"

The security guard he showed the card to looked like he might be a part-time biker with his shaved head and black goatee. He gave it an impassive look. "Sir, this is a private enterprise. We don't allow guns in Ghostland. It's not a constitutional thing. You want to start a petition, you go ahead and do that. But you can't walk in here with that weapon."

"Everything okay, Leonard?" Niko called over, puffing out his chest.

The bearded gun owner looked back and forth between the two giants and his shoulders slumped, far out of his weight class. He slipped the small—almost dainty—pistol out of his jogging pants and handed it over to the man Niko had called Leonard.

"Thank you, sir," Leonard said. "Now let's go get that paperwork filled out. You can pick up your *weapon*—" He spoke the word with obvious ridicule, holding the offending object between thumb and pinky finger. "—when you leave the premises."

The guard drew the retractable belt, snapping the gate closed to the groans of everyone in line, and ushered the gun owner toward the security kiosk.

Niko smiled and tucked the pills back into Ben's bag. "Go on in," he said with a wink. "You look like you got a strong heart."

"Thank you!" Ben took his bag back and hugged it excitedly to his chest.

"Welcome. Keep those pills handy, huh?"

"I will."

He caught up with Lil and Allison.

"All good?" Allison asked.

He nodded, slipping the backpack over his shoulders, relieved he'd made it through. If the guard had found his lighter fluid, he'd made up a cover story about forgetting it from when he'd gone camping last weekend, but the story hadn't felt believable even to himself.

Right past the gates, a large, colorful park directory stood above a stand filled with maps. He crossed to it and plucked a map out of a slot. They were folded in three sections. The front section was the middle of the park, with the Ghostland logo stamped above a cartoon archway. "You guys want one?"

Allison shook her head. Lil shrugged. "I'll just borrow yours," she said.

He unfolded the map and flipped through it. One of the features was called *Know Your Ghosts: A Guide to Ghostland.* This contained brief descriptions of several types of ghosts—Poltergeists, Possessors, Orbs, and a handful of others Ben had never heard of—along with the histories of a few park highlights. He folded it up and tucked it into his back pocket, eager to look it over when he got a chance.

"There's the house," Lil said, pointing to an exaggerated cartoon of Garrote House in the upper right corner of the directory. "Looks like we might have to go through the jail to get there."

"Fontaine County Correctional," Allison read off the map. "That name sounds familiar."

"It's supposed to be the most haunted prison in America," Lil said before Ben could chime in. He was surprised she still remembered. "Remember that stupid *Ghost Brothers* episode where they went there?" she asked with a big grin. "We used to laugh at those guys all the time, 'member? They always took themselves so fucking seriously."

Ben laughed. How many hours had they spent watching that show off the DVR before the days of binge-watching, he wondered. Too many to count, probably. He said, "I love how they were always going, 'Do you feel that cold spot?' or 'Did you guys just hear that?' Then they'd play back the footage and it'd just be some sound like the house creaking, but they'd put up subtitles like it was ghosts saying the stupidest random stuff."

Lil laughed. Then her expression darkened. "Back then we thought it was all fake."

"I still kinda think it was," Ben said, thinking of the Ghost Brothers specifically. But really any of those paranormal shows were all about the power of suggestion. "Just because we know ghosts are real now, that doesn't mean they weren't all faking it still."

They both realized at the same moment Allison was studying them like some kind of experiment. "Don't mind me," the therapist said. "Just Allison, remember?"

Lil rolled her eyes and turned back to the directory, visually mapping out the route with an index finger. "Looks like the fastest way to get there is through the Visitor Center. Then we can take this Ghost Tram thing to the midway and right to the prison." She popped her collar and turned to Ben with a supercool expression. "Let's go check out some ghosts," she said, mocking the dorky catchphrase the Ghost Brothers used to say on their show.

Ben laughed and followed her toward the park interior. As they passed between two metallic pillars on either side of the directory, a message appeared in blue digital typeface on the inside of Ben's glasses. He blinked and the words came into focus:

Exhibits within Ghostland pose no danger to our guests. However, if you believe you are being targeted by one of our exhibits, please contact Guest Services immediately.

PRESS Y TO CONTINUE

"Well, that's reassuring," Allison said.

Ben fumbled with his headset and pressed the leftmost button, which he remembered was the green Y. The first message was then replaced by another:

ONLY YOU CAN PREVENT SUICIDE!

If you see someone you believe might be in trouble, find one of our Suicide Prevention Officers in purple T-shirts immediately. You could save a life!

DO YOU UNDERSTAND YOUR OBLIGATION?
Y / X

"Suicide Prevention Officers?" Lil asked dubiously.

Allison said, "There was a story on the news the other night, something about a suicide challenge scheduled for opening day. Supposedly one of those flash mob things kids used to do, but the police said it was internet trolls fooling around. I suppose the prevention officers are to cover their butts from lawsuits, just in case."

Ben spotted one of them in the crowd, a young black man with a thousand-watt smile taking a selfie with the Japanese teens from the ticket booths. The guy looked to be only a few years older than Ben and Lil. He wore a purple T-shirt over his broad shoulders, with SPO printed on it in large white letters.

Ben wasn't sure how anyone was supposed to decide if someone was thinking about killing themselves. Even if he saw someone who looked depressed, he couldn't just snitch on them. It wasn't like reporting a crime or an abandoned piece of luggage. How could you prove someone was suicidal?

If I die in that house, that's what people will think about me, he thought. *That I was suicidal, or some crazy superfan, burning myself to death just like Garrote did.*

A third message appeared. He was used to them already. They no longer startled him.

HEADSET PAIRED WITH SYSTEM

"Awesome," Lil said. "Now what?"

As they approached a pair of tall monitors, Allison startled, then chuckled nervously and waved a hand in front of her face.

Rex Garrote[1] stood just a few feet away from her, not a ghost—at least Ben didn't think so—but a hologram. It was light years ahead of anything he'd experienced, including Twitch streams and walkthroughs of VR games he'd seen on YouTube. Comparing them to the man standing in front of them would be like pitting modern gaming graphics up against those old *Pong* machines. He could see individual pores on the writer's face, the tiny hairs on the mole beside his nose, the few flecks of green in his dark brown eyes, and the pulls on his trademark cardigan. This put some of the best CGI films to shame. If he hadn't known the man was dead, and if not for the slight shimmering quality to the image, the Garrote standing before him would have been indistinguishable from reality.

He almost felt like he could reach out and touch the man. He wondered if he *dared.*

Because looking at Rex Garrote right now, his former idol, all Ben wanted was to choke him until his eyes popped out of his head.

I hate you, he thought, struggling to hold back tears. *You ruined my life.*

The writer remained passive, devoid of emotions, an AI in mid-loop. Then the hologram glitched slightly and gestured toward the park interior. "Welcome to Ghostland, the most terrifying place on Earth," Garrote said. And with a knowing smirk, he recited the words he was most famous for, which had opened the old *Ghost World* anthology series: "Tell me... what are *you* afraid of?"

You, Ben thought. *You're still the scariest man in the world. But I'm gonna find you in that house, and I'm gonna burn it to the ground while you watch.*

He felt the bottom of his backpack, and the secret pocket the security guard hadn't noticed when he'd been distracted by the man with the tiny gun. The can of lighter fluid was still there. As were the Kitchen God matches, the same brand Garrote had used to immolate himself. Ben had planned to squeeze out the can on the old books in Garrote's private library, on the drapes and the carpets. He would leave the last few drops to spatter on his Garrote paperback, and use its brittle pages to set the house ablaze.

He smiled at the simple beauty of his plan.

And for the briefest moment, it seemed as though Rex Garrote's hologram smiled back.

KNOW YOUR GHOSTS

JUST PAST THE maps and the 3D screens featuring Rex Garrote's hologram, wide concrete steps led to the Visitor Center, a large, glass and chrome one-story building. Signs at the foot of the stairs said Start Your Interactive Tour Here!

The inside was noisy, with people milling around various exhibits. Not packed, but still a big crowd for ten in the morning. Already Lilian could feel a headache coming on. On both sides of the doors were racks of guidebooks. Ben grabbed one. Lilian picked one up herself. It was called *Know Your Ghosts*, written by the infamous Ghost Brothers, whose photos were on the cover with their black-leather-clad arms crossed over their chests, looking like wannabe rock stars.

"Friggin' doofuses," Lilian said. She laughed and slipped it back into its slot. The three of them moved into the crowd.

The interior was laid out like a museum, with exhibits on podiums or in cases or behind glass. TV monitors were stationed throughout: the monitor closest to the doors showed a time-lapse clip of the massive Fontaine County Correctional building being taken apart in large blocks by hundreds of workers and giant machines, then reassembled at Ghostland. She saw Garrote House hauled to its place on a hill, and the grass and trees and large stone gates placed around it in fast motion. Furniture was positioned in various exhibits, seats arranged in a darkened theater, a dingy circus tent going up, an old covered bridge lowered over a small manmade creek—all of this happened in the span of thirty seconds, with a timeline from 2012 to 2020 on the lower right corner.

They entered the SPIRITUALISM section. The first exhibit was titled "Spirit Rooms." Behind glass, candles illuminated a seance with several Victorian-era men and women seated at a table. The table began to wobble, then to rise and fall. It was obvious the medium was moving it with his knee. Suddenly a portly gentleman in a three-piece suit began to float, while the others watched in astonishment. Coins seemed to materialize from thin air, clattering to the table below. It wasn't until Lilian lowered her headset that she realized they weren't actors but part of the "interactive tour." Were they ghosts or just holograms? She didn't know, and the exhibit sign didn't specify.

"Neat," Ben said.

Lilian snorted laughter, but Ben had already moved on to a small group gathered in front of another glass case, this one on a podium. Inside was a strange-looking sailor doll on a wooden chair. Its face and feet resembled the cartoon monkey, Curious George. Below the case was a sign: *Robert the Enchanted Doll*[1]*, 1904* (Poltergeist) – *Owned by Robert Eugene Otto. On loan from the East Martello Museum.* Under this was a brief outline of the doll's history, which Lilian had very little interest in.

The small crowd jerked away from the glass, gasping and chuckling nervously. Ben joined them. As she approached the doll, it let out a high, squeaky laugh she heard in her headset earbuds. She thought it must use GPS location in addition to surround sound to be able to judge her position in regard to the exhibit, which she had to admit was pretty impressive.

Suddenly Robert the Doll stood and began to dance a jig. Its limbs flopped and its head lolled, but it didn't appear to have strings attached to it, and its movements were too jittery to be animatronic.

It's a ghost, Lilian thought. *A real ghost.*

She took a keen interest in the doll's history then, reading the words "unknown Bahamian girl" and "poltergeist" and "curse" as she skimmed the sign. Then she hurried to catch up to Ben and Allison, who had already moved on.

Another monitor showed an interview with Sara Jane Amblin. It was difficult for Lilian to hear the woman talk about her invention over the noise of the crowd, but Closed Captioning had been provided. "We wanted to create as authentic an experience as possible," the inventor said. "Ghostland was never just about frights and entertainment for me. First and foremost, it's an educational tool. A window into the past. A way for us as a society to look back at our collective mistakes and achievements and learn from—"

The topic didn't interest Lilian much. She'd hoped to hear—or read—about the technology. How it worked. How they'd trapped the ghosts and how they kept them contained. Again, Ben and Allison had moved ahead. She caught up to them at a large exhibit called "Pepper's Ghost," which showed the classic illusion of a ghostly figure projected into a scene using a mirror.

From there, they moved on to a glass case of old photos, labeled "Photographic Hoaxes." Most were portraits where gauzy apparitions appeared to hover near the heads of the subjects, created using double exposure, papier-mâché masks, coat hangers and cheese cloth. The majority of them looked like Halloween decorations made by kids. How anyone could have thought they were real made Lilian marvel at how far civilization had come in a little over a hundred years.

Beside these were modern photographs labeled "Orb Photography." In them, various subjects and settings—mostly cemeteries and basements—were marred by fuzzy circles, blobs and streaks. Lilian knew it was probably caused by dust on the lens, or within the camera itself. She'd taken a photography course last year, and learned about all kinds of old cameras. Some of the more recent snapshots in the collection were more believable. But even the so-called "Grey Lady" photo had been debunked as a botched iPhone panoramic, the "ghost" nothing more than motion blur.

In the "Mediums and Skeptics" exhibit they saw paintings of seances and "Mesmerism & Somnambulism," along with old books written by old dead people like Emanuel Swedenborg and Franz Mesmer. A diorama called *Seances in the White House?* displayed images of Abraham Lincoln and his family. Another cluster of images and articles was titled "The Ghost Club, 1862," which Lilian discovered by skimming the sign below had been a group of paranormal investigators in the mid-1800s, including the famous writers Charles Dickens and Sir Arthur Conan Doyle.

Alongside this was a diorama spotlighting Harry Houdini. Lilian knew him as a famous magician and escape artist, but she hadn't known he'd devoted much of his life to debunking spiritualists. There were photos of him demonstrating the trickery used in the Photographic Hoaxes and Spiritualism displays, along with posters for live shows and books with titles like *Houdini's Spirit Exposés* and "Do Spirits Return? Houdini Says No and

Proves It!"

As a skeptic since her mid-teens, the exhibit pleased Lilian. When she was little, she and Ben had been obsessed with ghosts and monsters and mysteries of the unknown. The two of them had often talked about traveling the world together, solving unexplained mysteries. When she'd grown older and wiser, she'd abandoned their childhood dream. At that point in her life, anything that couldn't be explained by science was a hoax. Eyewitness testimony was inherently faulty. Stories were often recanted. Like a young Dana Scully, she'd required hard proof, whereas Ben—the Mulder of their duo—had merely wanted to believe.

When the Amblin woman claimed scientific proof of the "spirit world," it had shaken Lilian. First Ben had been brought back from the brink of death, and now she had to contend with the fact that ghosts were real. The news had dropped like an atomic bomb. While grateful for proof of the existence of a human soul (if it could be called that—Sara Jane Amblin called it "dead energy"), many church groups resented the idea that spirits "lived" among the living, although Catholics and Lutherans began to claim it proved their long-held belief in purgatory or limbo. And while data-minded scientists demanded peer-reviewed papers, psychics and paranormal researchers had cheered the legitimization of their work, whether they were wide-eyed idealists or blatant frauds.

Suicides had spiked in the months following the news. The afterlife had to be better than this reality.

Even now, Lilian wasn't sure whether she believed in ghosts or not. She wondered what Houdini would have made of the discovery, whether he would have scoffed at the idea, or if it would have shaken him as much as it had her.

Whichever, she saw no proof here, only more holograms and hoaxes. And while the exhibits themselves were informative and entertaining, she wasn't about to be swayed by digital smoke and mirrors.

They approached a crowd surrounding an old car with round wheel coverings, circular headlights and white rims on the tires. Where the black metal wasn't riddled with bullet holes, it was in good condition, buffed to a high shine. Black smoke spewed out from the back. She could hear its engine rumble.

As they reached it, a man in a flat cap leaned out the driver's window, puffing on a cigar. He held a long gun with a flat cylinder she supposed must be where it held the bullets. Neither the cigar nor the exhaust had any smell, and Lilian realized they must be part of the "AR experience" promised to them out front. Even the sound of the engine was fake.

The exhibit sign said *Joe "Schmo" Russo*[2] (Apparition) – *Died Oklahoma City, 1935*. Joe Schmo cackled as he squeezed the trigger. A burst of flame shot from the barrel and phantom shells ejected into the gasping crowd. Still laughing, the gangster stepped out of the car without opening the door and began firing at the spectators. Several of them shied away, holding up their hands to protect themselves. A young boy started crying, slung over his mother's shoulder.

Then the gangster vanished, reappearing behind the wheel.

Ben said, "That was boss!"

"'Boss'?" Lilian sneered in Ben's direction. She didn't find it all that spectacular. It was basically a 3D movie without any plot.

If Ghostland intended to impress her, they would need to seriously up their game.

MANIACS

THERE WAS ALREADY a huge line at the Ghost Tram station by the time they got there. Lilian grumbled at the sight. At the front of the crowd, the Japanese goths wedged themselves into a tram car and it rose into the air. Twenty minutes later, Lilian and the others stood at the front of the line awaiting their turn. Passengers disembarked on the other side as the cars returned. A middle-aged couple wearing horror movie T-shirts and blue jeans hugged each other tightly, laughing nervously. The two dudebros with them sighed relief and ran their hands through their slicked-back hair as though they'd just been through Hell and back.

"Well, well, whaddaya know?"

Lilian recognized the man's voice before turning to see the retired detective standing behind her, wearing a smirk, fanning his sparse blond hair with the crumpled fedora.

Ben said, "Hey, it's the Map Man."

"What a pleasant surprise," Allison said with obvious sarcasm.

"We all enjoying the park so far, boys and ghouls?"

"Not as much as I thought I would," Ben said. "I just can't get over the fact these ghosts were real people."

Lilian agreed. On their way to the tram station they'd passed a handful of exhibits: a clock tower where a man shot at people in the crowd, a circus tent that had burned down with the show in full swing, a "hanging bridge" with ghosts swinging from the rafters. Each exhibit had been introduced by Rex Garrote's hologram with a pun-filled monologue worthy of his old TV series, each one creepier than the last.

Truth be told, Lilian wasn't sure how much more of this place she could stomach.

"Feels a bit like dancing on a mass grave, doesn't it?" Stan said, hunching up his shoulders and heaving a sigh.

Allison raised an eyebrow at him. "For once, you and I agree on something."

"Hey, that calls for a toast." Stan slipped a flask out of an inner coat pocket, twisted off the cap and raised it. "To good health."

"It's not even noon."

"Hey, do I tell you when to take your medicine?" He winked and swigged from the bottle. "You never did say what brought you here," he said, screwing the cap on his flask and slipping it back into his pocket.

"We came to see Garrote House," Ben said.

"Oh yeah? You know, I was the detective on that case."

"Seriously?" Ben said, his eyes practically twinkling with excitement. Lilian bet he would have loved to sit down and pick this guy's brain, if he had the time. He'd been inside Garrote House. He'd *investigated* the man's death. He probably knew things about Rex Garrote nobody else in the world was privy to. Ben was Rex Garrote's number one fan. He would definitely go nuts over it.

"Not one of my greatest accomplishments, but yeah," Stan said. "Dead serious, as the man himself might have said."

"Do you think it's true? That he set himself on fire?"

The detective chuckled. "I'll tell you what I think..." He cleared his throat and looked around like a man about to tell a dirty joke. "I don't think he's dead at all."

"You think he's still *alive*?" Lilian asked, skeptical.

"Sure. Easy enough to fake your death when you're swimming in money. I mean, look at this place. With all the cash his estate sunk into it, don't you think he'd have stuck around to see it completed?"

Ben frowned, obviously not buying it. "How could he stay hidden, though? Nobody's seen him for twenty years."

Stan shrugged. "False identity. Living out of the country. Nobody's *looking* for him, are they? I mean, if he was to pop up in Guatemala who would recognize him? Shave off that cookie duster, no one would know it was him. He's not exactly Elvis."

"But he *is* famous," Ben said.

"Oh yeah? How many writers do you actually know to look at? Could you point out Lee Child in a police lineup? How about Agatha Christie?"

Ben shrugged and muttered something Lilian didn't fully catch, something about not even knowing who they were.

"Exactly," Stan said. "I show a picture of Rex Garrote to the average person, they'll shrug like you just did. And I bet you dollars to donuts he's walking around this park of his right now pretending to be a goddamn hologram, just laughing his ass off. The guy's a maniac. I mean, just look at this place. What kind of ghoul would dream up a place like this?"

Ben looked around, shaking his head in awe. Like he was seeing the place in a whole new light. Or looking for Garrote to pop out and admit to them all the whole thing was an elaborate prank.

The group ahead of them rose into the air and the next car stopped alongside the platform. The woman working the tram opened the door and turned to them with a dead-toothed smile. She looked more like a carny than any of the previous employees with her green bug eyes and frizzy horsetail. "You're gonna wanna have your glasses on during the ride," she said. "You'll miss all the fun if you don't."

"What kind of fun?" Ben asked warily.

The tram operator winked. "Oh, all kinds," she said, and began cackling. She finished her spiel in a single breath: "Please-don't-try-to-open-the-door-once-you're-in-the-air-and-the-enjoy-your-ride-on-the-Ghost-Tram."

The three of them climbed into the empty car and dutifully put on their headsets.

"You coming?" Ben called out to Stan, who stood on the platform looking awkward.

"There's room for four," the tram operator said.

"If you insist." Stan took off his hat and ducked into the car as Lilian buckled her seatbelt. "Didn't want to impose," he said. He glanced at Allison, who shrugged as if she didn't care. But Lilian knew her therapist's mannerisms all too well. She was annoyed. And Lilian—who had tried so hard to annoy the woman in the past—couldn't help but feel a little pleased about it.

Rex Garrote's voice came over the tram's loudspeaker: "*Welcome to the Ghost Tram. Buckle your safety belts, boys and ghouls—*"

Stan said, "Hey, that's my line."

"*—for the most terrifying ride of your life.*"

Garrote quieted and for a moment they sat in silence, looking out the windows as the car rose into the sky, watching the rooftops become visible and the people below grow small.

"You know, Garrote's not the only maniac running wild in this park," Stan said.

Before he could continue, Garrote interrupted: "*If you look to your left, you will see Charles Manafort's pirate sloop* Gentleman of the Sea[1]*. And if you look closer, you may see the pirate and his crew, who'd perished upon the high seas in a fierce battle against the Royal Navy.*"

Ben stood and pointed out the window. "Check it out, you guys!"

Lilian looked over his shoulder as the tram car slowly passed the rotting sails of an ancient three-mast ship. Ethereal apparitions circled the masts like gulls, their ragged garments fluttering. One of them whisked past the window and Ben staggered back, almost bumping into Allison. Despite her fear, Lilian stood her ground. The ghost came back for another pass, so close she could make out his long straggly beard, gaunt face and haunted eyes.

"I bet those pirate ghosts are just Rex Garrote wearing a mask," she said, feeling pretty clever about the reference.

Ben grinned, catching on. "And he would've gotten away with it too—"

"—if not for us meddling kids!" they finished together and broke up laughing.

Allison looked at them quizzically and the retired detective grinned. He took his hat off again and started fanning himself with it. "Hot in here, ain't it?"

Lilian watched Ben stare open-mouthed out the window. Funny how quickly the two of them had fallen back into step after so many years apart. She had to admit it felt good. Familiar. It also scared the living hell out of her. She couldn't bear the thought of losing him again. It would hurt too much. It would leave a hole in her heart too big to fill.

"A retired cop is never really retired," Stan said to Allison when it seemed he wouldn't be interrupted again. "Too many cold cases to keep you awake at night. For me, at least. In particular—"

"*To your right is the infamous Apache Theater*[2]," Garrote said, causing the detective to pout at the interruption. "*This sprawling one-story theater was owned and operated by somewhat of a showman, who'd employed rather ingenious tactics to terrify his audiences. On this particular evening, each of the theater's seats was fitted with a 'shock chair' and wrist shackles for a special one-night-only screening of the cult film* House of the Zapper. *But their host was a madman bent on murder—*"

A crackle sounded over the speaker, cutting off Garrote's speech.

"I don't like the sound of that," Stan said.

Garrote's intro didn't resume. A moment later a frantic female voice shouted: "*—he's loose, dammit, one of those goddamn freaks got loose—!*"

Static swallowed her words. Lilian and the others looked around, growing troubled.

Allison said, "What was that all about?"

The car came to a sudden halt, slamming her against the window. Lilian fell against Ben. He caught her in his arms and chuckled nervously. "Sorry," he said. She broke the embrace. Ben was blushing.

"You think something's wrong with the ride?" Stan asked, looking nervously through the back window.

"I'm sure it'll start right up again soo—" Allison began, then shot a nervous glance out the window. "What was that?"

Ben followed her gaze. "What was what?"

Something thudded against the tram's metal exterior, reverberating like the inside of a drum, and the car rocked on its suspensions. The passengers grabbed onto something, looking around wildly in mounting fear.

Lilian held the handrail so tight her knuckles went white. They were too far up in the air. If the wire snapped, if the tram car fell, the impact against the concrete below would definitely crush them to death. She willed the intrusive morbid thoughts away, repeating a terror-filled mantra in her head: *Please don't let us die, please don't let us die.*

"Do you see anything?" she asked. Panic had made her whisper it.

Ben peered down through the window. "I don't see anything." He leaned closer to the glass. "Oh wow, it looks like we're right above the insane asylum."

"I had to go to Bright Falls one time to question one of the patients," Stan said. "I'm not ashamed to say that place scared the bejesus outta me, what with all the—"

Another violent drumming shook the car. Everyone held on to a rail or the bench except Stan, who fanned himself frantically with his hat. This time, it had sounded like footsteps. Lilian was certain of it. Like someone was standing on the roof.

The car behind them swung wildly, causing their own to wobble on the wire. Lilian could see the fearful expressions of its passengers, a family who looked like they belonged in an Old Navy commercial. The father, wife and children looked up as a man dressed in a pale green hospital gown and fuzzy slippers clambered up onto the roof, holding a cleaver.

Lilian jerked back in fright. How did he get up there? She blinked, stunned silent as the maniac ghost reached through the roof and grabbed the little blond boy by his hair. The child screamed soundlessly, his stubby little fingers scrabbling at the ghostly hand. The boy's parents grabbed him around the waist. The psycho kept pulling. If they didn't let go, they were going to tear him in half.

"Ben...?"

Lilian's voice sounded small even to her own ears. Somehow Ben heard her and turned from looking out the front. As his eyes widened in horror, everyone followed his gaze out the back.

"What the hell?" Stan said, pressing up against the window.

The child slipped free of his parents' grasp. His head slammed against the ceiling once, twice, before the ghost let go of his hair and he fell in a crumpled heap in the arms of his parents. As they busied themselves with the fate of their boy, the maniac dropped through the roof and began gleefully slashing away with his cleaver. Blood splattered the inside of the glass before Lilian could force herself to turn away.

"Jeepers creepers!" Stan said.

The father's anguished face mashed against the inside of the glass, streaking away a swath of blood like a human windshield wiper. He disappeared into the darkened car—blood had blotted out the windows, dimming the morning light—then slammed against the window again, mashing his nose and cracking the glass. Lilian could almost hear its high-pitched shatter.

Footsteps trampled on the roof above her head. She twisted to look up at the small emergency hatch. It locked from the inside, but she was acutely aware a lock wouldn't stop a ghost if it wanted in. It certainly hadn't stopped the mental patient from killing everyone in the car behind them, and if this goddamn tram didn't start moving soon, everyone in here would be the next to die, she was sure of it.

How did ghosts even get up here? Why aren't they in their exhibits?

The window of the car behind them shattered outward and the father's lifeless body came hurtling out of the car. In the jagged glass-framed hole, the killer stood drenched in blood, his knife dripping gore. In the same instant his face flashed—a glitch reminding her where they were and what these things must be—and for a split second, she was sure the man with the cleaver, smiling wide, his face painted with the dead family's blood, was the same man who'd welcomed them to the park.

Rex Garrote was watching her, smiling, his brown eyes gleaming with malevolent glee from the body of the psycho killer.

Lilian tore off her glasses. The earbuds came out with them. The ghost with the cleaver no longer stood in the window. The window wasn't even broken. The car behind them was entirely undamaged and free of blood. Even the screams had stopped. Another small group of spectators was staring out the back of the car, pointing fingers, screaming silently, gripping each other. Not the family they'd just watch get slaughtered.

They must have been seeing the same show in the car behind them, and likely the car ahead had seen a similar massacre in the car Lilian and the others sat in.

She laughed at her own foolishness and raised the glasses again just as a bloody hand reached down through the ceiling above her, blood-drenched fingers clutching feebly. As she lowered them again the hand disappeared.

"There's nothing there," she said. "It's a trick. It's all just part of the show!"

Ben took off his headset and blinked. He let out a relieved laugh at what he saw—or didn't see. Everyone did. It was obvious they all felt as foolish as she did. The tension vanished like a ghost without their glasses.

The tram shook again. Without the accompanying bang, it was no more frightening than a sudden gust of wind. Which was bad enough with her fear of heights, Lilian considered, but not worth freaking out over. Some mechanism had obviously caused it to move in time with the sound effects.

"He just Shyamalaned the hell out of us," Ben said.

"He? He who?" Allison asked.

"Rex Garrote," Ben said. "Who else?"

"What did I tell ya?" Stan grunted, fanning himself as he shook his head. "That maniac's laughing his goddamn ass off."

As THE CAR started rolling again, Lilian finally allowed herself to calm. The rest of the trip was uneventful. She did her best not to look out the windows, and tried not to pay attention to Garrote's sporadic narration of the exhibits below. Meanwhile, Stan told them the reason he'd come to Ghostland on opening day.

"Speaking of maniacs," he said, "I've been chipping away at some of the cases I was never able to solve. I figured this miracle technology—*the ability to see ghosts*," he said with sarcastic jazz hands, "—I figured they might try to use it to solve crimes, not to be the feature of a damn theme park. What did I know? I'm no scientist. Anyways, that's why I left the wife behind to drive all the way out from Emerald City for opening day. There was the Garrote case, of course, but I've made my peace with that one—as much as I'm able. No, the one I came to put to bed was the Doll's Head Murders[3]."

"What are the Doll's Head Murders?" Ben asked.

Eagerly, Stan told them about the case. Six women in the Seattle metropolitan area of a similar body type had each been bound and strangled to death, left with the head of a vintage fashion doll in their mouths, its little unblinking eyes staring outward, held between their perfect teeth.

Lilian said, "Do we have to talk about this?" She was already feeling ill, and talking about some sick twisted serial killer after what they'd just seen wasn't making it any better. She was eager to get back on safe ground. Her head was swimming from the height and the heat within the small, enclosed tram car.

"All right, I'll skip the gory details," Stan said. "I get a little carried away sometimes. Just ask my wife."

Stan explained his frustration with the investigation and how his young hothead partner had botched it by beating up their lead suspect—the son of a very rich and powerful woman—during questioning, forcing them to drop the case against Alexander Robin Fischer, the suspected Doll's Head Murderer.

Ten years later, one of the murder sites—a double-wide trailer salvaged from the defunct University Trailer Park—had been purchased by Ghostland, and since any close relatives of the victims were also deceased there had been no uproar. Stan suspected the Doll's Head Murderer would be here today to relive the murder through its recreation.

"That's likely," Allison said. "Pathological murderers often return to the scenes of their crimes, to reexperience feelings of power and often, sexual gratification."

"Gross," Lilian said.

Allison ignored her. "If one of his victims is on display here, in spirit... he might have come to gloat."

"That was my thought, exactly," Stan said. "He could have come in disguise but likely not. It's been ten years since the last murder. And like our writer friend, I doubt most people would recognize him to see him."

The retired detective then slipped a photograph out of his flask pocket and shared it around. Lilian glanced at it but found she couldn't focus. The man in the black-and-white photograph was handsome, well dressed and clean shaven. His eyes were dark and somewhat mysterious. She passed it quickly to Allison, who gave it a more thorough examination before handing it back.

"If you see this man, keep your distance," Stan said. "He's damn sure dangerous, and he may even be armed."

"What about you?" Ben said. "Are you armed?"

Stan drew a pair of handcuffs out of the breast pocket of his jacket. "These are the only defense I need." The cuffs clinked against the hollow flask as he slipped them back into his pocket and sat back with a contented smile. "I'm gonna bring that son of a bitch to justice today if it kills me."

Lilian didn't think it was likely. Looking at Stan, she thought it was more likely it would kill him, and she wondered what might happen to someone who died at Ghostland. Would their spirits move on, or would they be stuck here with the rest of them, unintended exhibits for customers to gawk at forever and ever?

She didn't like the idea of that at all.

Remind me not to die here, she thought.

The car swayed suddenly and the queasiness returned with a vengeance, driving a spike into her brain and making her gorge rise. While Ben and Stan began geeking out over the map Ben had kept folded into his back pocket, Lilian tried her best to keep from puking. And when they finally disembarked at the tram station, she stumbled out on shaky legs, rushed to the closest trashcan and threw up the coffee and half-digested toast she'd had for breakfast.

STALKER

BEN WAS STILL shaken by everything they'd experienced on the tram, but he'd been able to get through it without having to break down and take one of his heart pills. And despite his mixed feelings toward Garrote, he had to admit the whole thing had been a good bit of theater. Most people wouldn't have noticed the story Garrote had been telling right before the incident was a clue to what would happen, the one about the theater owner who'd murdered his audience with shock chairs. In retrospect Ben was upset he hadn't pieced it together before Lil had figured out the trick.

But it was Stan's murderer that was on his mind now as they left the tram station. The idea that real live murderer could be wandering the park set him slightly on edge, but it was also exciting. Stan had pointed out the exhibit he suspected the murderer would be heading for on the map, a building that housed haunted and cursed vehicles west of the curvy blue line marking the creek. It was where the detective headed now as Lilian raced toward the closest bathroom.

"I hope you catch your murderer," Ben said, waving.

"And I hope you find what you're looking for at that house," Stan said. "Sincerely, I do."

The detective tipped his hat and wandered off, leaving Ben alone with Allison. The two of them stood a moment in awkward silence. Finally, the therapist said, "It must have been difficult for you, to have gone through such a traumatic experience without a lifeline."

Ben frowned. "What do you mean?"

"You and Lilian used to be close..."

"Oh. That."

"If it's any consolation, she hasn't pushed you away because she doesn't like you."

"I know," he said. "It's because kids call me names. Make fun of me."

"Because they're afraid of you," Allison said, eyeing him as if to gauge his reaction.

"Why would anyone be afraid of me?"

"Because of what you represent. Teenagers think they're immortal. What happened to you reminds them they're not."

He shrugged. "I don't know. With all those school shootings—"

"Duck Falls is far enough removed from all of that. Trust me. They're afraid of you, Ben. Because you've lived through the most traumatic event a human being ever has to experience. Our common bond. Sooner or later, we all die." Allison looked around and chuckled darkly. "This place is proof of that, isn't it?"

"I guess it is," Ben said. He shuffled awkwardly, uncertain if Allison intended to continue the conversation or not. He thought about the implements of fire in his bag and wondered what Allison would think of him when his business was done. Would she tell the news reporters she'd suspected him all along? Would she tell them he was likely a chronic bedwetter and had practiced setting smaller fires before burning Garrote House to the ground, which he hadn't? Would she say it was always the quiet ones you had to look out for? The desperate loners?

After another long moment of silence Allison said, "When I was very young, I watched my grandmother die of Alzheimer's."

"Jeez, I'm sorry," Ben said. Her comment had come out of nowhere. The kneejerk platitude he'd offered never seemed sufficient and felt even less so now, after everything they'd seen.

The therapist smiled briefly. "Thank you. It traumatized me for many years, watching her lose her mind, wasting away to nothing, unable to do anything about it myself. Paper-thin skin draped over brittle bones. A decade of therapy later, I'm still somewhat numbed by the experience." She paused briefly, not leaving enough room for Ben to formulate a lame response. "That's why I wanted to come here with the two of you today," she continued. "I don't want Lilian to live through what I have. It picks away at you, trauma does. Like carrion on a corpse."

Lil emerged from the bathrooms. Allison flashed Ben a beseeching smile.

"Please don't tell her I spoke to you about this."

"I won't," he said.

Lil approached, her hair damp at the forehead and temples. She looked at them and scowled. "What? Do I have puke in my hair?"

"Nope, all good."

Her eyes narrowed. "Were you guys just talking about me?"

"No," he said hurriedly.

"Of course not, Lilian," her therapist said. "That would violate doctor-patient privilege."

Lil studied their faces a moment longer before nodding. "Good. Let's get going then. The faster we get to Garrote House the sooner we can go home." She hurried on ahead.

"I don't think I'm going to be able to put my glasses back on after what we saw on the tram," Allison said as they trailed along behind Lil.

"That was pretty messed up," Ben said. "But I'm kind of excited to see some more ghosts."

Beyond a small food court with a few of the typical theme park food stands—spiral fries and chicken fingers and candy apples and waffle cones—stood dozens of carnival games and rides—including a carousel and mid-sized Ferris wheel Ben was willing to bet were both haunted—along with several smaller exhibits set up down the middle of the road.

People walked the promenade, gawking and pointing, reading maps, plopping gobs of sticky pink and blue cotton candy into their mouths, pushing strollers and holding hands with lovers. A crowd had gathered around a gallows where a Ghostland employee in a silver Hazmat-type suit stood below the hangman's noose. The suit crackled with static electricity and the employee held up their arms as if to protect someone. Ben put his headset back on and saw several filthy Victorian-era ghosts with nooses hung from their snapped necks tormenting a muscular, hooded hangman. The employee in the protective suit—a sort of zookeeper of the dead, Ben guessed—stood between them, bursts of static erupting from the crinkly silver fabric with each ghostly fist that struck it.

"Whoa." Ben lowered the glasses. "I wonder what's up with that?"

"I don't know," Allison replied. "But the technology here certainly is *serious*."

Just beyond the gallows was a small building no larger than a trailer, where people waited in line to get their photo taken to look like a ghost. Ghostland employees directed attendees to individual photo booths, and above each one a monitor showed the finished products, which looked similar to those lenticular photos that changed depending on the viewing angle, the kind Ben used for Halloween decorations. The sign said "Ghost Your Selfie!" and promised to send the images as gifs to the attendee's email address.

"Neat," Ben said.

Nearby, a smaller crowd had gathered around a magic show, where a woman in chains drowned during a water escape trick and a magician emerged from a cabinet painted with the words THE MAGNIFICENT QUENTIN[1]. Without the glasses the water inside the empty tank splashed and the lock and chains rattled on their own.

Ben glanced around. "I think we lost Lil."

Allison, who stood half a foot taller than him, pointed through the crowd ahead. "There she is." As he followed alongside her, she turned to him. "She prefers Lilian, you know."

"We grew up together," Ben said. "She just prefers Lilian because calling her 'Lil' reminds her of when she used to be a geek like me."

"You don't strike me as a geek."

He gave her a stern look. "I'm home-schooled, my mom still kisses me when I leave the house and I read for fun. If that's not a geek, I don't know what is."

"I read for fun," Allison said.

"It's okay when you're an adult, I think."

She smiled. "Still, you should humor her. She'll respect you more."

"I'll try. It's just habit."

"Old habits die hard." She rose on her tiptoes and frowned. "Oh. She just ran into the funhouse."

"She loves funhouses," Ben said. "Wait, she *ran*?" He couldn't imagine why Lil would run into the funhouse without telling them where she was going. Could she be in trouble? Whatever the case, he picked up his pace through the crowd. "Come on, we'd better catch up."

LILIAN HAD JUST PASSED the drowned escape artist when someone shoved her from behind. She'd stumbled forward and almost lost her balance, scuffing the shell toe of her left Converse. When she turned to look, a few people in the crowd eyed her strangely but no one appeared to be the guilty party. Some rando had definitely shoved her, but that someone had likely already ghosted into the crowd.

Without a suspect, Lilian pressed forward. At this point all she wanted was to get in and out of Garrote House quickly so Ben could make his peace, and right now he and Allison were holding her up. She spotted them still at the hangman exhibit. Ben had his glasses on and Allison waited patiently at his side. She locked eyes with the woman for a brief moment, long enough for Dr. Wexler to smile in response to her scowl. Then someone pushed her again and she stumbled forward into the crowd, bumping into a large woman in a floral-print muumuu.

"Sorry," she muttered, looking around for her attacker. No one stood out. Again, a few random people gave her looks as they passed, and a kid eating nachos heaped with melted cheese and jalapenos giggled, wearing a fake cheese mustache. It could have been any one of them or none of them. Whoever it was, she couldn't imagine why a stranger would choose her to pick on of all people. Did she look like an easy target? Was she wearing a sign that said *Kick me*?

She kept walking, glancing over her shoulder every so often, even though she kept telling herself to ignore it. Whoever it was would get bored and go pick on someone else. She spotted a massive building with a domed glass roof behind a stone wall at the far end of the fairway and recognized it as the prison. According to the map at the entrance, Garrote House wasn't far beyond that, which meant she would only have to suffer through two more exhibits before they could get the hell out of here.

She literally could not wait to put this day behind her, although she was kind of glad she'd allowed herself to relax around Ben. He was a good kid. Strange but fun, in a dorky sort of way. They'd had a lot of good times together in the past, and she felt sincerely bad about having ducked him all this time. Before the day was through, she promised herself she would apologize. But first they had to get to Garrote House and then get the hell out of this place.

A carnival funhouse stood between Lilian and the prison, multicolored lights spelling out ROCKY'S FUN WORLD[2] on its wacky-looking sign. The exit was a spinning tunnel with disorienting spirals painted on the inside, nestled between the legs of a giant gorilla taller than the building itself, which slowly pounded its chest as it swayed from side to side.

Lilian had a ton of fun memories of funhouses. Her favorite was always the mirror mazes and the wobbly stairs. The only things she didn't like were the fat mirrors and the circus music, and it blasted from the speakers loud enough to wake the dead, although at least this song had a kind of jazzy sound to it.

Something struck her right foot and she went sprawling, scraping her hands and knees on the concrete. "What the hell!" she shouted up from the ground at the people passing by, most of them oblivious. "Whoever the fuck keeps messing with me, you'd better stop it right now!"

A cute boy in a Letterman jacket reached out to help her up. Lilian slapped his hand away.

"You think it's real funny pushing girls around? You think that's gonna get you laid?"

The boy looked at her like she'd gone crazy. His dark pompadour flopped animatedly as he spoke, a strand of hair falling out of place. "I don't know what you're talking about, man. You just tripped."

"I didn't trip," she snapped back. "*You* tripped me."

"I was watching you, man," he said with a sneer. "You tripped over your own feet. I thought maybe you had like epilepsy or somethin' but I guess you're just nuts."

"I'm not nuts," she called after him as he walked away shaking his head. "*You're* nuts!" She shot to her feet, muttering, "Stupid asshole." Her knees felt bruised and where her jeans had been fashionably ripped, she saw that she'd skinned them. Her palms burned, the skin raw and bleeding lightly in places. She brought her right hand close to her face to tweeze a small stone from an open wound with her fingernails, sucking a breath through her teeth.

Someone grabbed her wrist and made her slap her own forehead so hard her vision filled with stars. Stunned, she gaped open-mouthed at her hand. She peered around wildly, trying not to whimper, angry and afraid all at once, searching the crowd for her attacker as she unconsciously thumbed the beads on her bracelet.

"What's her problem?" the muumuu woman said, passing by with a sour expression.

Lilian slipped a badly shaking hand into the lapel pocket of her jean jacket and took out the glasses. Her hands trembled as she put them on.

The maniac from the asylum stood before her, panting and grinning through a mask of blood, so close she could see the burst capillaries in his zombified eyes, the various shades of brown and yellow in his rotting teeth and the frayed stitches in the scar along his forehead. Slowly, he raised the dripping cleaver above his head. Lilian staggered back as the blade arced down, cutting through the space she left.

"*Leave me alone!*"

The killer shuffled forward, slippers scuffing on the concrete, blood and drool spilling from his lips. As the surging crowd gawked at her like she was some sort of carnival freak, she realized nobody else could see him. And if she didn't run away *right now*, she was dead.

She bolted, pushing through the opposing bodies, spinning and rebounding her way through the crowd until she reached the entrance to the funhouse. With a single glance over her shoulder, she dashed up the aluminum steps, hoping like hell she could lose the freak inside the house of mirrors.

ESCAPE

BEN PAUSED AT the entrance to Rocky's Fun World, wondering if the thirty-foot-tall gorilla towering over the exit tunnel was supposed to be Rocky or what. The exhibit sign set him straight: Rocky had been the owner of the traveling midway in which this funhouse had been featured, not the gorilla. Rocky Arnault had killed himself by cutting pieces off of his body in his own mirror maze until he'd bled to death.

Allison caught up to Ben at the entrance and brushed past him, uninterested in the funhouse's history, the soles of her dock shoes clanking rapidly up the metal stairs. Her anxiety put Ben on edge. Earlier she'd told him it looked like Lil had been chased into the funhouse but she hadn't seen anyone run in after her. Whatever had happened, it wasn't like Lil to run off the way she had, even considering how often she'd personally ghosted him in the past. If she'd been planning to go into the funhouse, she would have told them.

He followed Allison into a mirrored hallway that seemed to stretch off into infinity. Angles had been painted on the pillars and the floor tiles were marked off with phosphorescent triangles. The lighting switched between normal, blacklight and darkness, the last of which illuminated only the triangles on the floor, while "Baby Elephant Walk" boomed over the sound system. Laughter drifted from deep inside the maze.

He'd never been great with mazes. Lilian had always taken the lead when the opportunity came up in games. She had a natural instinct for them, and seemed to enjoy possessing a skill he didn't himself, often lording it over him. Calling him a "noob," with a sardonic smile he'd been able to sense over the headset. Back then he would never have been able to imagine how much he would miss those times. How he'd wonder what she was up to while he played their favorite games with strangers, kids who called games "vidya" now and every other word was "rekt" or a personal attack.

Thinking this, he bumped face-first into a glass wall and Allison crashed into him. "Sorry," he said.

"My fault," she said.

He slipped past her and headed down a different path. He had to keep his mind on the maze. Lilian was in here somewhere. If someone had chased her, she could be in danger. She needed their help.

He paused, feeling for glass to his left and right like a mime in a box. To the immediate left was a mirror, its surface flat and cold. His reflection stared back at him, the fear in his eyes amplifying it, making it sharp as broken glass. He turned right.

"Mirror," a guy somewhere up ahead of them said. "That's a mirror."

A moment later a young couple walked by holding hands. They jumped back in fright and cursed at something at their feet. Ben saw the severed hand and forearm, lying limp against its reflection like a strange plant or a table lamp in a pool of tacky blood. The couple walked away from the sight, heading deeper into the maze.

Ben headed toward them, hands held out. His wrists jammed up hard against cold glass. Another dead end.

Come on, Ben, concentrate!

The couple was behind them now. There were two more mirrors to his left and his right. Stood against the right was a foot, still within its scuffed and paint-stained work boot, severed above the ankle. He jerked back in horror, expecting to see Rocky's ghost in the mirror when he looked up. But there was only his reflection, with Allison looking at her own reflection over his shoulder. She looked as confused and uneasy as he felt.

"I've never been much for carnival rides," she said with a nervous chuckle.

"We'll find her," he said. He sidled by her and followed the young couple, hoping it wasn't their reflection. They screamed and hugged each other, likely spotting another piece of Rocky Arnault's ghost. Ben turned right, left, then left again, avoiding mirrors and glass, trying to steer clear of dead ends and the bloody appendages he knew had been left there, all the while glancing behind himself to make sure he hadn't lost Allison. He didn't trust just seeing her reflection, even if she was right at his feet. Something about the maze made him think he couldn't trust his own eyes.

"*Leave me alone!*"

He stopped and listened. Allison gave him a quizzical look, her mouth fixed in a grimace of anxiety. It was hard to hear much over the music, the laughter and screaming but he was pretty sure he recognized the voice. "I think I just heard Lil."

"That scream?"

He nodded.

"We have to find her," Allison said, her eyes big with worry.

The lights went out again, illuminating only the tile markings. Ben stumbled ahead, hands in front, palms forward. He touched glass. In the dark he couldn't tell if it was a window or a mirror. He felt along the frame for an opening and stepped through.

Glass to the left and right. He moved forward another tile. Glass again to the left and right. A severed ear at his feet, a splash of blood on the glass. Cold sweat dripped down his ribs from his armpits and his heartbeat quickened. It wasn't quite bad enough to worry about, not yet, but he'd have to keep a handle on his breathing or he'd need to take a pill. The pills made him dizzy, groggy and unalert. Sometimes they gave him migraines so bad he'd have to lie down in the dark with a cold cloth on his forehead. He couldn't allow for that. He needed to be wide awake and focused.

Lil needed him.

When the blacklight came back on, Allison was right behind him and the mirrors on either side stretched their reflections into infinity. Ben turned back and slipped past her, heading down another corridor. He turned right again, following number one of Lil's unwritten rules for escaping mazes: *turn right whenever possible*.

At the end of the corridor, a set of janky stairs led up to the next level, where a man with no legs hovered, his mouth a lipless grimace, his lidless eyes forever staring, a blood-drenched hacksaw gripped in his remaining hand.

BEN WATCHED the young couple step out of a passage at the foot of the stairs. They screamed at the sight of Rocky Arnault's ghost and the man slashed at them with the hacksaw. Shying away from it, the couple turned in Ben and Allison's direction, their outstretched hands pressed flat against glass, going white as the blood fled escaped their capillaries. The guy said, "Mirror!" The woman hugged him and they turned around. Ben hurried forward and his nose slammed into glass. Allison almost bumped into him. She looked around in confusion and vague irritation.

"There they are," she said, pointing toward the moving staircase. She headed for the stairs. With her glasses off she couldn't see the ghost hovering there.

"Be careful," he said. "He's right there."

Allison scowled back at him. "They're harmless," she said. "Lilian needs our help."

Ignoring the warning, Allison stopped directly in front of the ghost. It eyed her with curiosity, cocking its head to watch as she stepped onto the first of the herky-jerky stairs. She grabbed the handrails and began her cautious, wobbly ascent, while Rocky slashed at her with the bloody hacksaw, its jagged-toothed blade swishing through her. Allison didn't even notice.

Steeling himself, Ben stepped through the ghost and onto the first stair. Each step titled and jerked mechanically, up and down and side to side, jostling them as they ascended. He had to keep his hands flat against the walls so he wouldn't stumble.

The ghost was still watching them from the bottom of the stairs, his body parts pieced back together like Frankenstein's monster. Ben raced past Allison, who held the railing in a death-grip, seemingly unsure where to step next as the stair she was on rose up and down and the next two moved side to side in opposing directions.

Ben flew over them, his feet barely touching the ground. Allison could handle herself. Lil needed him. Nothing else mattered now.

He pushed through heavy, clear plastic-strip curtains like the kind found in meat lockers—covered in visible fingerprints and likely sweat and saliva, possibly even a few stray boogers—and stepped out onto a balcony lined with convex funhouse mirrors. From the far end he heard Lil scream, "*Why are you doing this?*" over the rumble of the moving staircase.

In the midway below, people milled around eating junk food, laughing, playing carnival games and listlessly weaving their way through the crowd. The giant gorilla swung out his furry forearms, momentarily blocking out the sun as it slowly beat its chest. "Baby Elephant Walk" looped again from the speakers, close and loud.

Lil screamed again, behind a thick black curtain leading into the gorilla's chest. He hurried past the mirrors, not caring how he might look as a fatter or taller kid and certainly not shorter or skinnier, his shoes clanking on the platform grating. He pushed through the black fabric and hurried into a small room, its walls painted luminescent red to look like the chambers of a heart. They seemed to grow and contract in time with a heartbeat pounding over the loudspeakers.

His own heart beat rapidly. *Got to calm down*, he thought. But when he spotted Lil crouched in the far corner, hugging her legs and rocking forward and back, it only made him feel worse. He hurried to her side. Tears glistened on her cheeks as she stared out into space, like she didn't see him there at all. She was counting the beads on her bracelet again. Her headset lay broken at her feet.

"Oh my God!" Allison gasped behind them.

Ben put a hand on Lil's shoulder. She flinched and her gaze darted toward him. "Ben... is that you? Is that *really you*?"

He nodded cautiously. "Yeah. It's me," he said, thinking, *Who else would it be, Lil? What happened to you?*

She touched his face and drew her hand back quickly, startled, like she'd expected to touch cold glass or for her hand to go right through him, like she thought he might be a reflection... or a ghost.

"What happened, Lilian? Are you okay?"

"*We have to go*," she hissed, her eyes darting around the throbbing chambers of the heart, the darkened corners, her face bathed in its blood-red glow.

"Okay, we can go now," he told her, much more calmly than he felt. "There's probably only one or two more rooms—"

"*No.*" Her gaze flicked toward the black curtain leading out. "Not out of *here,* out of this *whole place*! Out of Ghostland! There's something *wrong* here, Ben."

She fixed him with a dead-serious stare, the fear in her eyes unlike anything he'd seen outside of a movie. He half expected to find, once they left this room and its angry red lighting, that her hair had gone partly white like it often did in old black-and-white horror movies to people who'd borne witness to Things that Should Not Be Seen and Could Not Be Unseen.

"Something is *really wrong*!" she cried, snatching out to grab him by the shoulders. He took her hands gently and placed them in her lap as Allison knelt down opposite him.

"Lilian, everything's okay," the therapist said. "We're going to get you out of here."

"Dr. Wexler?" Her eyes flicked toward the woman.

"I'm here."

Ben hadn't ever heard the woman speak so softly, so much like a therapist. She slipped a hand under Lil's elbow.

"Come on, sweetie. You're safe now."

"*Am I?*"

"Yes. I promise you are." She helped Lil to her feet. Lil swooned and placed a hand against the wall to steady herself, while Allison took much of her weight.

"You have to stay with me, Ben," Lil told him.

He swallowed hard and said, "I will."

"*Promise me!*"

"I promise I won't leave you, Lilian," he said, picking up her headset and tucking it into his backpack where it settled against the Garrote book. He tried to smile for her but the lie made it twitch on his lips.

She nodded, causing a tear to run a shimmery trail down her cheek. Allison led her out of the giant's beating heart. Ben followed them, disappointment weighing him down.

It was entirely possible something bad really was happening here, like Lil said. He knew that. He could almost feel it himself.

But Garrote House still called to him, more desperately now than ever. Particularly if there was something dangerous in this place. He had to go there, he had to find the man behind the curtain, pulling the strings. He'd never be able to convince Lil to stay though, not after this. He supposed he'd have to come back some other time, on his own if it came down to it, but once his parents discovered he was gone today it would be even more difficult to get out of the house next time. They would probably keep watch on him twenty-four-seven. Even if he could manage to get back here, it wasn't likely to be so easy to get through security with the lighter fluid, the matches, his pills. He'd be busted for sure next time.

The three of them staggered out through the spinning vortex, dazed and on edge, and while Allison comforted Lil, Ben looked up at the giant gorilla beating its chest and wondered what the hell had happened to her in there.

"Lil... Lilian," he said. Allison was smoothing Lilian's hair. "Were you wearing your glasses inside the funhouse?"

Lilian's head jerked in his direction.

"They were broken," he said, pulling them out of his bag to show her. "Did you drop them?"

She nodded. "I h-had to," she said. Her teeth chattered. "He was following me. I couldn't see him without them but he was... he was *stalking* me."

Allison looked around cautiously. "Who was stalking you, Lilian?"

"The ghost. With the c-cleaver."

"From the tram?" Ben asked, hoping she hadn't heard the doubt in his voice. He'd expected her to say "Rocky," to tell them she'd been chased by the limbless ghost, dripping with gore as it floated or crawled after her through the maze. To hear her say she'd been followed all the way out here by the ghost from the tram... he wasn't sure if he could vouch for her sanity anymore.

Allison raised an eyebrow at him. "Lilian, that ghost couldn't have followed you. It's just a hologram. A small part of a computer program."

"I know what I saw!" Lilian's lower lip quivered as fresh tears spilled from her eyes. "He kept *pushing* me. I ran... I ran in here to hide from him but he *followed me in.* I don't care if it's not possible, that's what happened—I know it's what happened! *He was trying to kill me!*"

Allison hugged Lilian against her chest. "Okay, shh. We're here with you now. It can't hurt you. Whoever or whatever it was can't hurt you anymore."

Lilian shook her head, her eyes squeezed shut. Ben studied her expression. He knew she didn't believe Allison and he wasn't sure he believed her either. She didn't just seem to be messing around to get attention or to make sure they wouldn't stay. He would have called her out if he'd thought so. Whatever had happened to her in there, Lilian *actually believed* she'd been targeted.

Harassed by a ghost.

Haunted.

Something was badly wrong, just like she'd said.

Lilian Roth was losing her mind.

Or Ghostland is real, he thought.

Whichever it was, Lil needed help. She was cracking apart—worse than the day he'd woken up after surgery to find her sitting at his bedside in the hospital. And he refused to stand by and watch that happen again.

BREATHER

ONCE THEY'D GOTTEN Lilian back out into the midway Allison pulled Ben aside. "This is exactly what I was worried about," she told him in a hushed tone.

He eyed Lilian to make sure she wasn't listening in. She was still toying with her bracelet, staring vacantly into the crowd. Tears had streaked her mascara. She'd wiped it off on her sleeves but the result had given her eyes a haunted, skeletal look.

"What do you mean?" he asked.

Allison shot Lilian a look of concern. "I think she may be experiencing vivid hallucinations caused by prolonged exposure to the AR glasses or the Recurrence Field technology."

"What if she's telling the truth?" Ben wasn't sure he believed Lil himself, but the alternative was worse. If anyone could start hallucinating at any given moment this place was headed toward a major disaster.

"Either way, she needs help," Allison said. "And I don't think we're capable of providing it, knowing as little as we do about the technology. They must have done trials, double-blind studies." She shook her head in uncertainty, adding, "Something like that."

Ben considered it. He tried to recall the message that had appeared on the inside of his glasses when they'd first entered the park. Other than leaving Ghostland—which he still didn't want to do, not before doing what he'd come here for—it was the only thing he could think of to help her.

Not the Suicide Prevention Officers, he thought, *hopefully it won't ever come to that*. But suddenly he remembered and blurted out, "Guest Services!"

Allison gave him a strange look.

"Remember the message? 'If you feel like you're being targeted by a ghost go to Guest Services right away.' Or something like that."

Lilian had returned to their side and gave Allison a hopeful look. She'd been listening in, at least to the tail end of their conversation.

The therapist nodded. "Let's see if we can find one."

Ben got the map out of his back pocket and unfolded it. The legend had a section for SERVICES with different icons for Food (a pizza slice), Washrooms (little toilets), First Aid (the Swiss cross), and Security (a badge). He found the little orange question mark for Guest Services at the western edge of a snaking blue line that would likely be Conococheague Creek and traced the way back with a finger. From where they stood at the end of the midway, they would have to cross the creek on what appeared to be a covered bridge and go past Ghost Town, USA[1]—a Wild West town marked with an ominous cartoon skull and crossbones wearing a black cowboy hat.

"Okay, I can get us there," he said. Lilian eased a little, enough to smile and relax the hunch in her shoulders, but not enough to give up on twirling her bracelet. "It's much closer than the entrance, and we won't have to go back on the tram. Looks like it shouldn't take long. It's too bad we can't stop at Ghost Town, USA, though. I heard they have this huge shootout every half hour. It sounded pretty cool from the ads."

"Maybe next time," Lilian said, though he doubted she was serious.

"You think you'll be okay?" Allison asked her. "I can hold your hand if you like."

"Don't be extra."

"Extra what?" Allison asked.

"I said *don't* be extra."

Still got her sarcasm, Ben thought. That was a good sign, at least.

"How about if I keep watch?" he suggested. Lilian and Allison gave him uncertain looks, so he explained: "I'll put on my headset so you don't have to. That way if the creep from the tram comes back—" He nearly added *or anyone else*, for some reason thinking of Rex Garrote, but he caught himself just in time. "—then I can warn you."

Lilian considered it a moment. She wiped her tears with the sleeve of her jean jacket and nodded.

Ben smiled at her and put on his glasses. She returned the smile warily and the three of them set off toward the bridge.

The other visitors shrieked and gasped and laughed at exhibits, everyone having a great time, while Lilian sulked and Ben stalked ahead, blazing a trail through the crowd.

Even though he wasn't sure he believed she'd actually seen the maniac from the tram, he kept a wary eye out for it. But apart from the ghosts held securely within their exhibits, caged between the Tesla poles like animals at the zoo, he didn't see any at all.

LILIAN STILL COULDN'T RECALL anything that had happened to her inside the funhouse. She remembered running inside, fleeing from the knife-wielding maniac, and then she was outside again, reunited with Ben and Allison, streaks of mascara itching her cheeks, her hands and knees bleeding and bruised. She was sure the lobotomized monster had found her in the maze and attacked her again, but she couldn't remember how she'd managed to escape him or how Ben and Allison had gotten her out of there. The rest was a blur of blacklight flashes and warped reflections.

Her stomach began to rumble as they passed through a sweet-smelling cloud from the nearby ice cream stand. The scorched-sugar tang of freshly made waffle cones and real vanilla. She craved chocolate, but anything to soak up the churning stomach acid would sure "hit the spot," as her dad often said after supper. "Guys, can we stop? I'm starving."

Ben plopped down into a molded plastic seat. He swung his backpack onto the table—it made a hollow metallic clunk as his water bottle or whatever hit the plastic tabletop—and he unzipped it, rummaged for a second and pulled out a granola bar. There were only a few people seated at the other tables, eating ice cream, banana splits and hot dogs. They ate mechanically and spoke in hushed voices, if at all, as if the horrors they'd been through getting this far had traumatized them as much as Lilian felt now.

"I guess we could stand to take a break," Allison said. "I need to use the facilities."

Lilian sneered at the word "facilities," and while her therapist headed off toward the washrooms, she slipped a few folded bills out of the tiny front pocket of her jeans and approached the ice cream stand. She waited her turn, keeping watch for the maniac from the tram even though she wasn't wearing her headset, merely watching the faces of the others gathered by the food court to gauge their reactions. Because if a ghost happened to wander into their midst while they ate their snacks and early lunches, Lilian wouldn't be the only one worrying.

Ben chewed his granola bar and glanced around at the wandering crowd. Despite how he'd likely led Allison through the mirror maze to get her, Lilian, to safety, it was obvious his heart was no longer in it. It wasn't fair to make him continue while she and Allison remained blissfully unaware. She would have to suck it up and put her own glasses back on soon. Ben needed his partner back. Someone to watch his six for a change.

The woman in the truck handed her an ice cream sandwich and Lilian paid for it, peeling off the plastic and biting into it as she returned to the table to sit opposite Ben.

"It's okay to be scared," she told him.

"I'm not scared," he muttered.

"*Ben*. We're all scared. I'm shitting myself right now, okay? Look, I know you're supposed to be desensitized to this stuff. I get it, you watch a lot of horror. Maybe you've just got too much empathy for a place like this."

He shrugged, looking off.

"Ben, you're *allowed* to be scared. You don't have to be my rock. Remember back in the fifth grade when Brody Lobban used to come over to my place after school sometimes and he'd always want to look at Rotten.com and Bestgore and sick stuff like that?"

Ben nodded. It was one of the reasons they'd stopped hanging out with Brody even though he was one of the only other kids who was as big a horror fan as the two of them. His morbid fascination with *real* death and *real* gore —the way he'd laugh and cringe at the same time while scrolling through one macabre photo after another—had made Lilian wonder if there hadn't been something wrong with him psychologically.

"What happened in the funhouse, it was worse than that," she said. "*Way* worse."

With a shudder against a sudden chill, she forced herself to focus on something else. She watched the woman in the truck window shape heaping scoops of orange and black ice cream onto a large waffle cone for a child far too small to eat that much.

"I don't want to think about what he might do if he finds me again," she said. She looked down at her treat, dripping vanilla ice cream down her thumb. She'd only taken a few bites and already the thought of eating any more made her want to puke again. "Do you want this?"

"Yeah, okay." He took it from her, licked the side of it and wolfed it down in three bites while Lilian wiped the melted ice cream off her hand with a wet wipe from his backpack.

"You don't always have to be the hero, okay? That's all I wanted to say. We're in this together, you and me. Dr. Wexler—*Allison*—she's a tough chick. And in spite of what you saw at the funhouse, I can be pretty tough too."

"I know," Ben said. He looked up from crumpling the ice cream bar wrapper and held her gaze. "You're the toughest girl I know."

She smiled. He smiled back. "I'm a boss bitch," she said with a grin.

Ben laughed. "I wouldn't say 'bitch,' but... kinda, yeah."

She laughed with him.

Allison returned from the washrooms and gave them a curious look. "You two are in a cheery mood."

"We just needed a Snickers," Ben said.

Allison smiled but didn't seem to get the joke. "Good. Shall we press on?"

He zipped up his backpack and stood. "I'm ready if you are."

Lilian stood alongside him. "Are my glasses still in there?" He nodded and she held out her hand. "Give 'em here."

"You sure?" he asked with a look of uncertainty.

Lilian said, "We have to look out for each other, don't we? You got my six, I got yours."

He shrugged and nodded, opened his backpack and handed her the glasses. One of the lenses had cracked but they looked like they still might work. She slipped them on, blinked to adjust to the crack slightly blurring her vision, then flicked them back on. After a moment a message popped up on the undamaged right lens:

REBOOTING

Once the progress bar had reached 100% another message appeared, with ellipses to indicate loading: *dot-dot-dot, dot-dot-dot.*

SEARCHING FOR SYSTEM...

Finally, it was replaced by:

HEADSET PAIRED WITH SYSTEM

Allison put her headset back on too. "All for one and one for all," she cheered halfheartedly.

Lilian clapped her hands together with far more enthusiasm than she felt. "Damn right," she said. "Now let's go check out some ghosts!"

Everyone laughed. Lilian took off her jacket and tied it around her waist as the sun came out from under a cloud, and the three of them left the food court in strangely good spirits.

VIRTUAL INSANITY

GUEST SERVICES WASN'T too far from the food court. Ben had pointed it out a short distance ahead after an almost five-minute walk to the southwest. Lilian glanced at the time on her cell phone. It wasn't getting any bars. No Wi-Fi, either. Sometimes the service was spotty on the outskirts of town, but usually it would flip back and forth between no service and full bars. This was just flat, like she'd flicked on airplane mode. It felt strange to be so completely disconnected from the world, even if she had no time at the moment for the internet.

"Hey, I heard they borrowed Annabelle for that exhibit," Ben said as they passed a low cinderblock building. The sign out front called it the Museum of Cursed Objects[1]. "The original Raggedy Ann doll, not the stupid one from the movies. The haunted mirror from Myrtles Plantation too. And they even got the original Dybbuk box."

He sounded honestly disappointed to be missing it. She had to stifle a chuckle.

A few minutes later, they reached Guest Services. Lilian stepped into the small, air-conditioned building first, followed so closely by Allison it felt like her therapist was a helicopter hovering over her shoulder. Directly inside was a small waiting room and a counter with a few computer terminals. TVs behind it showed crossfading images of Ghostland's many exhibits along with the video of Rex Garrote from the park entrance on repeat, like a sales pitch in a travel agency.

Lilian headed straight for the counter and began tapping her fingernails on it, until the plump woman with a fluffy blond fringe and big brown doe eyes looked up from her computer terminal and did a double-take. The woman stood abruptly, her large breasts stretching out the red Ghostland tee she wore, and hurried around the desk. She plastered on a huge smile and said, "Hiiiiiii! Oh my goodness, what happened to your *hands*, baby girl? Are you okay?"

If it hadn't been for the intermittent dull burn shooting up her wrists, Lilian might have forgotten all about the injuries she'd sustained from the pavement in the midway. It was her knees that had taken the worst of the fall, and they still throbbed badly. She held her hands up to look at them for the first time since she'd left the funhouse. Rivulets of blood had seeped down, dried and cracking in the folds of her palms and between her fingers. "They're okay," she said. "Actually, is there anyone we can talk to about—"

"She's been harassed," Allison said, speaking over her.

Lilian glared at her until Allison noticed and offered an apologetic shrug.

"Oh my!" The Ghostland employee had the sort of chubby, rosy-cheeked face which only seemed to jibe with expressions of cheer or shock. The look of dismay seemed foreign, squishing up her features. "Honey, we take allegations of sexual harassment very seriously—"

Lilian felt her cheeks grow as red with embarrassment as Ben's might. "It wasn't sexual!" she cried. "*Gawd!*"

"It was a ghost," Ben explained, looking embarrassed himself. "And it wasn't... it wasn't sexual," he finished unenthusiastically, then gave Lilian a pained, quizzical look as if to ask, *Right?*

The woman said, "Oh." She sounded almost disappointed, letting her shoulders slump for a moment. She glanced over her shoulder toward a closed door. "Let me just see if Demont is done his break."

"Can't you help me?"

"Demont is one of our SPOs," the woman said hurriedly, already heading toward the back room. "He's far more qualified to deal with situations like this."

"She's not suicidal," Ben said.

"I'm not suicidal. He pushed me." Lilian held out her hands palms up and demonstrated.

"Pushed," the woman said dubiously. "By one of our ghosts?"

Lilian didn't like the woman's tone but other than storming out and tweeting a complaint once she got back within network range, there was little she could do about it. Besides, they'd already come this far. She needed somebody to take her seriously. She hoped this Demont guy would be the one.

"He was a mental patient, I think," she told the employee. "He was wearing a green gown and had a scar across his forehead."

"A prefrontal lobotomy scar," Allison said.

"And he had a cleaver," Lilian added, annoyed by Allison trying to speak for her again. "He followed us from the tram ride."

"Which was pretty cool, actually," Ben said. Lilian gave him a look that wiped the smile off his face and he shrugged. "What? It was."

"After the ride, I went ahead on my own," she said, ignoring Ben. She didn't want to dredge up what had happened to her inside the funhouse, but she knew talking about the maniac was the only way to get these people to take her seriously. "Somebody pushed me. I thought it was just some random in the crowd but when it happened again, I turned around and saw... *him*." She swallowed hard, on the verge of tears just thinking about him standing over her, blood-drenched and drooling. "He swung at me with the knife and that's when I ran into the funhouse."

"We found her in one of the rooms practically catatonic," Allison said.

"And you believed her," the woman said. It wasn't a question.

"Why *wouldn't* she believe me?" Lilian snapped.

"I'm her therapist. I've been treating Lilian for over a year, and I have no reason whatsoever to believe she'd be lying to me, nor do I believe she's been hallucinating. If she says she was harassed by one of your—whatever these things are—I believe you people need to take that seriously."

The Ghostland employee exhaled heavily through her nose. "Okay," she practically moaned. "It's just that we've never had to deal with something like this before. I mean, there were the stories during the construction..." The woman raised a hand to her mouth as if to cram the words back inside, but she clearly knew it was too late.

"Something happened during construction?" Ben asked.

"Let me get Demont," she said, continuing to the door. "Don't go anywhere."

Allison said, "We won't."

As the door closed behind the woman, Lilian turned to Allison. "I don't think they're going to believe me."

"What happened during the construction?" Ben asked again.

"*Who cares what happened during the construction?*"

He shrank away from her. "Well, if someone died or got injured," he said, his voice small, his head nearly swallowed by his shoulders, "don't you think that might explain what happened to you? What if someone got attacked like you did?"

Lilian thought it over. If he was right, they were talking about multiple incidents. That would mean there was a much bigger problem here than singling her out. Before she could respond the woman returned with a young black man in a purple T-shirt, slim with broad shoulders. He was still chewing whatever he'd been eating on his lunch break and wiped his hands together to rid them of crumbs. He stuck one out as he approached her, ran his tongue over his teeth and smiled. "Hi, I'm Demont. And you are?"

She shook his hand, recognizing him from when they'd first entered the park, where he'd been taking selfies with visitors. "I'm Lilian. This is Allison. That's Ben."

"Hi," Demont said, waving at them casually. "Have a seat." He directed them to the waiting room chairs where they all sat down, looking anxious. He took one of the chairs, flipped it around backwards and straddled it like a dancer in an '80s music video with his forearms draped across the back. "So," he said finally, holding direct eye contact with Lilian until she felt herself blush. "You met Morton."

"Morton?" Allison asked.

He nodded. "Morton Welles[2]. A psychiatric patient at Bright Falls Sanitarium in the early-1900s. Violent psychopath, revenant-type haunting. And very dangerous outside of his Recurrence Field loop, I'd imagine."

"So what was he doing on the tram ride?" Ben asked.

"Well, Ben..." Demont turned to him. "Legend has it, Mr. Garrote thought it would be fun to put a little jump scare on the Ghost Tram. I personally don't like jump scares. I think they're cheap. But if you've ever seen his TV show, you'd know he was a big fan."

Ben nodded as if he knew very well, which of course he did.

"Morton has taken a liking to women around your age before," Demont continued. Lilian took note of the term "woman," and tried not to let her blush deepen. "I guess you'd say it's his M.O. When the park was still being built, one of the student psych volunteers—a white girl around your age, dark hair with freckles—she disappeared."

"*Disappeared?*" Allison said.

"Just for a few hours. When they found her again, in the operating theater of the asylum, she was... *disturbed*, I guess." He gave them a brief, unhappy smile. "She described pretty much what you experienced. Said she'd been pushed. Had her hair pulled. Then he swung the cleaver at her. I don't want to put more of a scare to you than you've already got but when they found her, she had slashes on her arms." He held out his left forearm, palm up and made slashing gestures with his right hand. "Where she claimed Morton had cut her," he added.

Ben muttered, "Jesus."

Demont turned to him briefly. "I don't think Jesus was anywhere near the asylum that day." His comforting gaze returned to Lilian. "This woman was under the impression she was only able to escape Morton Welles because of where she hid. She said he wouldn't cross into the operating room, for some reason. That he stayed outside the doors looking in through the windows until help arrived."

"They would have performed his lobotomy there," Allison suggested.

Demont nodded without looking away from Lilian. "That's what I thought too. Most people dismissed her. Said she'd made the cuts herself, that it was a cry for help. I've never believed that, personally. I've always thought something did come after her. And now he's come after you. A full-bodied revenant. Morton Welles. You saw him with or without your glasses?"

"With," Lilian said.

"That's good. If he'd shown himself to you without the glasses that would have been something further to worry about."

Allison shook her head in frustration. "So you've had violent incidents in the past and they still opened this place to the public?"

"Like I said, nobody believed her. As far as park officials are concerned, this place is one-hundred percent safe."

"People thought that about the Titanic."

Demont gave her a weary smile. "You're absolutely right." He stood, spun the chair around and pushed it back into place. "And I think it's time we go see the wizard," he said.

Lilian shook her head in confusion. "The wizard?"

Demont nodded. "Sara Jane Amblin. If there's a problem with the Recurrence Field, Miss Amblin needs to know about it. And sooner, rather than later, I'd say." He turned to the blonde woman as she reemerged from the back. "Lola, do me a favor: call in to the control room, tell them we're on our way to see the boss lady. This young woman's been harassed by one of our ghosts."

The woman nodded, bug-eyed with apparent fright. "I'll call right away." She turned abruptly, bumped into the desk, and stumbled on her way to the backroom.

Demont frowned briefly, then ushered the three of them to the door.

While the others went on ahead, Lilian watched Rex Garrote repeat his walk toward the camera on the television screens, each a half-second out of step with the other. She was close enough to read his lips as he asked: "...what are you afraid of?"

You, she thought. *I'm afraid of you, Rex Garrote.*

For a moment she considered telling them she'd seen Morton Welles change, that right after murdering the family in the tram he'd transformed into Rex Garrote. But the murder had been part of the show. If she told them what she thought she'd seen, they would begin to question her story. They would question her *sanity*.

Better to not say anything, she thought.

On the televisions, Rex Garrote smiled as if he agreed.

A SILENCE HAD fallen as the Suicide Prevention Officer drove them along the thoroughfare in a red Ghostland golf cart, honking its tiny-sounding horn here and there at pedestrians. Ben watched the exhibits go by. Lilian stared at the back of his seat. It was Allison, sitting in the back with Lil, who finally broke the silence.

"What made you want to work in a place like this?"

Demont glanced over his shoulder, seemingly surprised by the question. "I'm a BSW student. Taking my Masters in Social Work in the fall. Ghostland was looking for student help and I figured I could use the experience."

"Then you have no issue with its content?"

"Its content?" Demont scowled slightly, his grip tightening on the wheel. "You've been listening to the GRP2 folks, haven't you?"

The comment seemed to offend Allison. "My opinions are my own, thank you," she said.

"Of course. I didn't mean to imply otherwise. Personally, I'm not the biggest fan of its presentation. A lot of the exhibits are... well, exploitative, to say the least. But I do think what Ghostland is doing is important, even if they may be doing it in a way in which I don't necessarily approve."

Allison arched an eyebrow. "How so?"

"As in 'those who forget the past are doomed to repeat it.'"

Demont glanced at Ben before returning his gaze to the path ahead. They passed an old, rambling one-story cinema, APACHE THEATER spelled out in red lights on the marquee. Two showings tonight: *The Shining* at 7PM and

House of the Zapper at 9:30. Ben recalled the story Garrote had told during the tram ride. *Zapper* was the movie the theater owner had rigged to electrocute his entire audience. Ben hadn't seen the film but he'd of course watched *The Shining* multiple times, the classic Stanley Kubrick film based on the Stephen King novel. He'd even seen the miniseries once or twice, and while he loved the movie, he thought the miniseries was truer to the book.

"Care to elaborate?" Allison asked.

"Think about World War II," Demont said. "Those death camps in Poland. An anti-Semite or a Holocaust denier strolls into Auschwitz, maybe thinking about how Hitler's 'master plan' wasn't all that bad. Then he sees that room full of hair shaved off the prisoner's heads. Another one full of glasses. Another with shoes. Maybe then it starts to hit him how truly awful the place was. Now, imagine he's standing there among all the prisoners, watching them starve, watching them being gassed—"

"Can we *not* talk about this?" Lilian said from the back.

"Sorry." He drummed his fingers on the wheel a moment, contemplating. "What I mean is, would he start to feel sympathy toward them? Any sane person would. Personally, I believe bigotry is a form of temporary insanity." He smiled at the notion. "Sometimes not so temporary, unfortunately. But it *is* something that can be overcome. That's been proven. It's learned behavior. Like cult indoctrination. In a way, Ghostland could be considered cultural deprogramming," he added with a chuckle.

Allison hummed thoughtfully, much louder than she would have if the powerful little electric engine wasn't whining so loudly. "I'm not sure I fully agree," she said. "But it is an interesting theory."

"Hey, believe what you want. I'm an optimist, personally."

"I'm a realist," Allison said.

As the two of them continued their debate, Bright Falls Sanitarium[1], a red brick building with multiple towers and green weathered copper roofs, stretched on and on to their left. Ben tried to imagine the operating theater Demont had mentioned, where the maniac Morton Welles had been lobotomized, and the student had hidden from him. It made perfect sense that Welles wouldn't have wanted to go into that room. If he'd been lobotomized there, as Allison suggested, he would have wanted to avoid the scene of his mutilation. Thinking this, Ben peered over his shoulder and saw Lilian glowering at the building.

"Are we almost there?" she asked, her eyes narrowing.

"Just about," Demont said.

They passed what looked like a Civil War fort on the right, followed by a massive pool containing the galleon they'd seen from the tram, rising and falling on artificial waves. Ben looked up at the ragged, flapping sails to see if he could spot a pirate ghost, but the plastic roof of the golf cart blocked his view. There were strange shapes rising from the water, which from the distance looked more like hands than waves. He turned to get a better view and Allison caught his eye, giving him an inquisitive look. He forced a smile, eager to appear compos mentis. He didn't need to be under her scrutiny any more than he was already.

Demont turned the cart to the left and pulled up alongside what looked like a squat government building, two stories constructed of cinderblock with no windows and a single door at the far left that looked like it might be made out of steel. There was a single camera bubble above the door and one on each corner of the building. Several of those metallic poles had been placed every few feet around its perimeter.

It all seemed like overkill to Ben but he supposed they must have been meant to keep ghosts out rather than in. Which got him wondering just how safe the people who made this park really thought it was that they would feel the necessity to build themselves an impenetrable fortress right in the middle of it.

THEY ALL PILED out of the golf cart and headed for the building. Demont approached the door, sorting through the set of keys on his belt. He selected one and tried it. It didn't fit so he tried the next. As the second key slid home he grinned. "There's the sweet spot," he muttered, twisting it in the lock.

The bolt made a heavy clunk as it unlatched and the door gasped open as if it had an airtight seal, revealing a long dim hallway with cinderblock walls beyond.

"Ladies first," Demont said, holding the door.

"And people say chivalry is dead," Lilian grumbled.

Demont gave her a wide smile. "No better place to resurrect it."

She rolled her eyes and stepped through the door. Allison went next, thanking him. Ben followed. Demont pulled the door shut behind them and locked it. It made the hiss of a tight seal. He flicked a switch on the wall and the overhead lights buzzed on one after another, lighting a path down the narrow corridor. The air smelled like cold concrete and ozone.

"Probably best if I go first from here on out," he said, tucking the keyring back into his pocket as he slipped between Allison and Lilian.

"Oh, we 'ladies' are just the canaries in the coal mine," Lilian said.

Demont laughed as they followed him down the hall. "There's nothing dangerous in here, Lilian. The big boss just doesn't like strangers playing in her sandbox. See the cameras?" He pointed to another tinted plastic bubble above them and a second where the hall veered left. "This building is the safest place in the park."

"Is that because of the poles outside?" Ben asked.

Demont winked back at him. "That's part of it."

"How do those things work, anyway?"

"I'll leave that to Miss Amblin. She likes to talk about her inventions."

Ben followed close behind as Demont turned the corner and headed down another corridor without a door. At the end the hall they took another left turn.

"Is it just me, or are we going around in circles?" Lilian asked. She was touching the outer wall, the way she did in mazes. Ben could see she was becoming increasingly aggravated with each left turn. They were violating her cardinal maze-running rule, but at least they were keeping to the outside. Otherwise she might have wigged out.

"Another good catch," Demont said. "You guys are on fire."

The next hall ended at a cinderblock wall. To its immediate left stood another metal door. Demont approached it and pressed a button on the intercom. The small speaker crackled. "Who goes there?" The woman's voice echoed in the narrow hall.

Demont waved up at the camera and thumbed the intercom button. "It's me, Miss Amblin. Demont Hudson. I'm a Suicide Prevention Officer. I've brought some people I think you need to speak to."

After an overly long pause, she asked, "To what is this pertaining?"

"Well, Miss Amblin, it seems one of our ghosts has been harassing this young woman—"

The door buzzed before he could finish and the Maglock disengaged with a clunk. He pulled the door open and ushered them into a large room reminiscent of NASA's Mission Control Center. Two full stories tall, its far wall was taken up entirely by security monitors. Ben thought there had to be at least fifty of them, rotating through various views and angles of the park, showing visitors walking through the exhibits, screaming, laughing, eating, everyone having a good time. He saw the tram ride and the funhouse, the asylum, the theater, the circus, cameras both inside and out. The security monitors must have been equipped with the same technology as the headsets, because in every shot the ghosts were just as visible as the guests.

There were only a handful of employees seated at computer terminals, far less than the desks available, which led in rows toward the monitor wall. The rest of them must have been at lunch.

Sara Jane Amblin approached the four of them down the aisle between the desks. She wore a black suit jacket and pencil skirt and the tightness of her face made her look harried, like the interruption was taking up her precious time. The wall of monitors reflected on the lenses of her thick-rimmed glasses.

"So," she said brusquely, folding her hands at her waist as she turned to face Lilian. "You believe one of our apparitions has been bothering you, is that correct?"

"I don't *believe* anything," Lilian said, sticking out her chin. "It happened."

"Hmm," the woman said with a thoughtful nod.

"Apparently, he was very physical with her," Demont said. "Pushing her. Pulling her hair. Just like what happened with Parker. He tried to cut her—"

"Nothing was ever proven in that instance," the inventor said.

"Maybe not. But two strangers come forward with basically the same story about the same ghost, I'm willing to give them the benefit of the doubt."

Sara Jane Amblin stared off at the wall of monitors, her eyes narrowing. After a moment she turned rigidly and asked Lilian, "Are you hurt?"

Lilian shook her head.

"Well, that's a relief, at least." The inventor looked thoughtfully at her hands, still clasped at her waist. "Perhaps we should consider deactivating him."

"*Deactivate*?" Ben said. "You can just shut them down like that?"

The woman studied him a moment. "I suppose since you've gotten a peek behind the curtain it wouldn't be so awful to divulge a few trade secrets." She flashed a brief smile that didn't reach her striking dark eyes. "Technically, what we'd be doing is removing several lines of code in the computer program which holds them in a form of stasis. Although Welles was one of Rex's favorites, it would be a shame to lose him."

Allison looked confused. "Hang on... *are* they ghosts or are they all just holograms?"

Sara Jane gave her a patient smile. "Oh, I assure you they are very real. You see, wherever we go our bodies leave waste energy behind, something like car exhaust. The most *volatile* waste energy is released when a person passes on. All of that waste energy—or *dead energy*, as we like to call it—has to go somewhere. In a sense, death is like an oil spill. Some deaths, especially *violent* deaths, are more like nuclear meltdowns. These are what laypeople think of as hauntings. My Recurrence Field technology traps this dead energy and amplifies it. This room, all of these computers, they're what make the so-called 'ghosts' visible to guests through your Augmented Reality headsets."

"So, they're ghosts *and* holograms," Ben said, thinking he understood.

The inventor nodded. "In a way. The tech is similar to holography, the difference being that holography uses reflected *light*, whereas what we like to call *spectography* tracks these wandering pockets of dead energy and synthesizes their analogs on our computer array."

Ben nodded, finding it difficult to follow but not wanting to appear dull.

"Of course, our digital effects team has taken liberties in certain cases where the imprints aren't quite as observable and the photographs and history are lacking. In essence, every single apparition in this park has a *digital* imprint as well as their energy imprint—you can think of the ghosts you see as avatars of the dead energy they've left behind. The Recurrence Field uses that digital imprint to physically contain them within a fixed loop, repeating the same actions and events ad infinitum. Or whatever we choose to make them do. The Ghostland program controls these loops and sends three-dimensional avatars to your headsets, but since both the avatars and your AR headsets are often required to interact with their environment, the main program also has some control over the physical tech in the exhibits, like doors, lighting, and various moving objects. Does that make sense?"

"I told you she likes to explain it," Demont said with a smile.

She gave him a look of annoyance.

"S-sorry, Miss Amblin."

"This all sounds very... exploitative," Allison said.

The inventor's attention snapped toward her. "I assure you it's not. With respect, these aren't sentient beings we're discussing. Without the tech we've created here they would simply float aimlessly, like sightless fish forever bumping into the glass of their tanks."

"Okay," Lilian said, "then how did Morton Welles chase me from the tram?"

"Morton Welles is what we call a 'free-roamer,'" the inventor explained. "Garrote insisted we have several of them. The Recurrence Field keeps Welles connected to the Ghost Tram show and away from his natural habitat, so to speak—the Bright Falls Sanitarium—which he would naturally gravitate toward without the restrictions. I think he likes the power trip, frankly." She rubbed her chin in thought. "But there's nothing *preventing* him from roaming beyond the tram. No ESP—"

"Extra-sensory perception," Ben said.

"Actually, it stands for 'ElectroStatic Precipitator,'" the inventor corrected him, making him feel dumb for having spoken up. "Those poles outside the building, surrounding the exhibits, they generate negative ions—a well-known repellent of dead energy. You may have heard some cultures use salt to ward off spirits? Once I concluded it was the negative ions present in the salt that appeared to prevent dead energy from crossing these barriers, it was a short mental leap to employing ESPs for the same purpose. Lines of salt all over the park just wouldn't be feasible or very stable, would they?"

Ben parsed this. "So basically, you've got a bunch of evil ghosts floating around the park like dogs off their leash."

"Only a matter of time before one of them bites," Allison added with a scowl of disapproval.

"They're not off-leash, per se. It might seem... problematic, to say the least, but it was a stipulation in Mr. Garrote's will, and the Hedgewood Foundation was adamant his wishes be adhered to. Why he'd picked some of the most violent apparitions to be free-roamers has always been a bit of a mystery, but our exhibits are completely safe, and this is the first time any of them have acted out—"

"Second," Demont reminded her.

She nodded, as if she were making mental calculations, tallying up the damage. "We should have deactivated him after the first time."

"And it's only opening day," Allison said. "Do you expect the odds to improve over time, or get worse?"

The inventor glowered at her. "I'm not a statistician."

"And I'm not a gambler," Allison replied with a satisfied smile.

The two women eyed each other for a moment. Ben wondered which would flinch first.

"Okay, Garrote be damned," the inventor said finally. "You've made up my mind. I'm shutting Welles down." She crossed to a balding man who sat in front of a computer station. "Harrison, bring up the code for Morton Welles, please."

The man looked up with uncertainty, the monitors reflecting off his smudged and dandruff-flecked glasses.

"Well? Go on."

"Mr. Garrote isn't gonna like that," he said, his voice nasal as if he had a cold.

"Mr. Garrote doesn't *have* to like it. He's been dead nearly twenty years."

Ben thought, *Not according to Detective Beadle.*

The programmer—Harrison, she'd said—brought up the program. Under PASSWORD REQUIRED he banged out a flurry of keystrokes and hit Okay. The code appeared and he scrolled through lines filled with underscores and hashtags and parentheses and words Ben knew but couldn't understand in their current context. The man clicked a short-key that brought up a bar labeled BOOLEAN SEARCH. He typed "name = morton_welles + object code" and pressed Enter.

More code appeared with "morton_welles" highlighted multiple times. The programmer scrolled again, paused at a line, then took off his glasses and rubbed them on his shirt. When he put them back on, he blinked hard at the screen. "That's not possible," he said.

Sara Jane leaned over his shoulder. "What's not possible?"

"The coding has changed." He pointed. "Here, this line should have an *X* but it's got an *N*. And this line should say 'type equals revenant' not 'type equals all.' This is bad, this is very, very—" He stood abruptly, scowling out at the rows of computers between himself and the monitor wall. "Has someone been tampering with the main system coding?"

FOR A MOMENT NO ONE SPOKE, eyeing each other accusatorily like characters at the end of an old detective movie. Ben, who stood watching the computer over the programmer's shoulder, saw several characters in the middle of the screen change. The programmer sat back down, fuming. "There!" he said, pointing at the changes Ben had already spotted. "Another line just changed! *Who's logged into the system?*"

"There's only the five of us in here, Harrison," a young woman at one of the computers up front said. She wore a Pokémon T-shirt and had her short, dulled pink hair pulled up in a small ponytail at the top of her head.

"Is it possible someone's using remote access?" Sara Jane asked.

The programmer shook his head. "The coding isn't accessible outside of this room. Hedgewood made very specific demands—"

"Not even to hackers?"

"I'm telling you, *unless it's one of us*—" He stood and peered around the control center accusingly. The Pokémon fan shook her head in exasperation. "—this is just not possible, ma'am."

"Everyone please, step away from your terminals," Sara Jane said, broadcasting her voice to fill the room. The handful of operators stood and reluctantly stepped away from their desks. Sara Jane turned back to Harrison. "Well?"

The programmer took off his glasses and shook his head, pinching the bridge of his nose. "It's still changing. It's almost like—"

"A virus," Ben said.

"Like a virus," Harrison agreed, nodding.

Sara Jane gave him a serious look. "*Could* it be?"

"It's possible, sure. But to what end?"

"Sabotage? One of our competitors? The GRP2 wackos?"

Harrison slipped his glasses back on. "I suppose. I just don't see how anyone could have uploaded it without direct access codes, and only a handful of us have them."

"Do we trust those people?"

"With my life? Not likely. But with the source code..." He nodded, pinching the bridge of his nose again. "Yes, I believe everyone in possession of those access codes has only the project in mind. However it is this happened, we're gonna have to reboot the system to flush it."

"How long would that take?"

"Hours. Depends on how deep the virus has gotten and how malignant it —"

Harrison launched out of his chair abruptly, causing Ben to leap out of his way and the chair to strike the desk behind him, where it toppled. "*Shit!*" the programmer gasped. He covered his mouth and stared fearfully at the screen.

"I don't like the sound of that," Demont said.

Harrison wagged a finger at the computer. "I *knew* I recognized that code. That right there—" He pointed at the screen in two separate places. "That's Rex Garrote's algorithm."

Sara Jane frowned. "What do you mean?"

"I mean Garrote's core algorithm has almost entirely copied over the algorithm for Morton Welles." He ran a hand through his sparse blond hair. "This isn't a *virus*, Miss Amblin." Jabbing a finger toward his terminal, he hissed, "*That's Rex Garrote's AI.*"

"Double shit," Sara Jane muttered. "How could this happen?"

"Honestly, ma'am? I haven't got a clue. All I know is, we have to shut everything down before the Garrote code replicates throughout the entire system."

"How dangerous could that be?"

"With the Recurrence Field and the ESPs still up and running, I'd say not very. But I'm not one-hundred percent sure."

The inventor nodded. "Best to be safe then." She reached for the intercom microphone nearby and twisted it toward her. "Attention, all patrons," she said, causing people on the security monitors to react in real-time. She released the Talk button and cleared her throat. "This is a general safety announcement. Exhibits will be shut down for a brief period due to routine computer maintenance. During this time, we ask that you please follow staff members to the exits in a calm and orderly fashion. Exhibits should resume shortly, however if you are unable to wait, the cost of your tickets will be refunded at the front gate. Thank you for your patience." She sighed heavily, releasing the mic. "That should buy us some time. If the three of you would like a refund or perhaps a gift certificate..."

Allison said, "I just want to forget this day ever happened."

Sara Jane nodded, again looking down at her clasped hands. "Okay. Shut it down, Harrison. You have one hour."

"I need more time, ma'am."

"*One hour*, Harrison."

"All right. I'll do what I can." The programmer righted his chair and rolled it over to the terminal. As he began typing furiously, Sara Jane turned to Demont, her expression grave.

"What the hell are we going to do in the meantime?" she asked. "Forty dollars a head. Those people out there will *crucify* us."

Demont gave her a sympathetic look. "I could rally up the troops. Let them know there might be some angry people looking to cause trouble."

"That would be great. Demont, was it?" He nodded. She put a hand on his shoulder and squeezed it lightly. "Thank you, Demont."

With his ego visibly inflated, Demont headed for the door, where he turned back. "It was nice to meet you all," he said with a small wave in Ben and Lilian's direction.

"You too," Lilian said. "Thank you."

"My pleasure. Take care of yourself, Lilian. You too, Ben."

"Thanks," Ben said.

"We will," Lilian added.

"Good luck with your schooling," Allison said with a wave. Demont thanked her and left. The door clanged shut, the lock engaged with a click and the seal hissed tight.

"Should we...?" Allison hesitated. "Do you want us to go?"

Sara Jane turned from watching over the programmer's shoulder and blinked, as if she'd already forgotten they were there. "Stay. By all means. You've seen the worst of it."

They stood around for a few minutes, watching Sara Jane and Harrison work. Ben wanted to learn more about Ghostland, but Sara Jane hovered over the programmer's shoulder, pointing out lines of code and asking about them. Finally, she stepped back and let the programmer do his thing, which seemed to relieve Harrison, and Ben took the opportunity to fish.

"This place is pretty amazing," he said.

She smiled briefly. "Thank you. It's a labor of love, really. I've been fascinated by paranormal phenomena since I was very young." The inventor seemed to lapse into reminiscence before flashing back to the present. "Fortunately, with the help of Garrote's estate and the Hedgewood Foundation, I've been able to achieve what no one in history has before. Someday, I hope to be able to use this technology to further our understanding of what lies beyond the veil, as they say. For now, I'm pleased with what we've been able to accomplish here."

Allison said, "It's very impressive, I have to admit."

The inventor thanked her.

"We came to see Garrote House," Ben said. He was wary of showing his hand but he wanted to know more about Garrote's involvement in the park. He wanted to know if this Garrote code was volatile or benign. Mostly, he wanted to know if the holograms of Rex Garrote that popped up around the park were the same sort of "free roamers" she had mentioned in reference to the mental patient Morton Welles. Could he show up anywhere? Could he be multiple places at one time?

"Oh?" the inventor said.

Ben nodded. "When the house came through town four years ago, I saw Mr. Garrote in one of the windows. I knew it was him because I'm one of his biggest fans. I've read all of his books and I even have the 25th anniversary Blu-ray of the series—"

"And you say you saw him in that house?" the inventor asked, scowling.

"Yep. Uh-huh."

"Without AR glasses." Her tone was dubious.

"Yeah. Should that not have happened?"

"To say the least," Sara Jane said. "No one has seen Rex Garrote in that house since he passed on. *Ever*. Not with the glasses, not without them. Not before we brought it here or after. As far as my team has been able to ascertain, Rex Garrote's dead energy is not present in that house. It never was. His image—his *source coding*—has all been replicated from old video footage, digitally enhanced."

"Then how could I see him?" Ben asked, wondering if Stan had been right, if Garrote really had faked his own death after all. It would explain why they hadn't found the writer's dead energy in Garrote House. It would also explain why he'd seen the man in the window that day when no one else had before or after.

Because he really is alive, Ben thought. *And he's here now, somewhere, watching us like ants in a terrarium.*

"You didn't see him," Sara Jane replied.

"I know what I saw."

"It was a hallucination."

"Bullshit," Lilian said.

The inventor looked at her as if she'd been slapped. "Excuse me?"

"I said *bullshit.* Ben saw him in that house. I know this for a fact. He had a *heart attack* because of it. I was on headset with him when it happened. He said 'I think I saw Rex Garrote,' and then his heart stopped beating. That doesn't happen when you hallucinate a friggin' ghost. If Ben says he saw him in that house, that's what happened. How else would he have known it was his house?"

Ben smiled and thanked her.

"Regardless," Sara Jane said dismissively, "whatever you saw or think you saw, it doesn't matter. The Rex Garrote you've seen at Ghostland isn't an apparition. It isn't a poltergeist or a multiplier or an elemental or any of the other types of entities. It's a computer algorithm and that's all it is. Ones and zeroes in the shape of a man."

"But it doesn't make a difference, does it?" Ben asked. "You just said you use the same algorithms for all the dead energies in your exhibits. Even if you just made Garrote out of old footage or clothes stuffed with newspaper, if he's a part of that program he's still just as real as any of the ghosts are."

Harrison stopped typing briefly and turned back from the computer. "The kid makes a good point. Actually, he's possibly *more* real than the others since he's partly artificial intelligence, like the free-roamers."

"But there *is* no dead energy!" Sara Jane cried, throwing up her hands in exasperation.

"Well, you see, there is," Harrison said, "because now he's taken over *actual ghosts* using your invention and my coding." He cleared his throat and pushed up his glasses before adding apologetically, "Ma'am."

Sara Jane narrowed her eyes a moment, then jabbed a finger at his computer screen. "Just shut that thing down."

"It's taking longer than expected. I had to close some external software,"

Harrison explained. The screen went black. "There we go."

The computer rebooted with a musical chime.

"Hey!" the pink-haired woman shouted—Nia, Sara Jane had called her—as she turned in her chair. "My computer just shut off." A heavyset guy in a Deadpool hoodie said, "Mine's out, too." A moment later, the whole wall of monitors suddenly went black all at once.

"*What the hell is happening?*" Sara Jane demanded.

"I don't know, ma'am." Harrison pushed his glasses up the bridge of his nose with a middle finger. "The reboot should have only affected terminals running the Ghostland program..."

The security monitors all came back on at once, mass pandemonium displayed on each one.

They watched with increasing horror as crowds ran screaming in silence. As guests were slashed, flung and burned, limbs and heads torn from bodies. The ghosts had escaped their exhibits and these "sightless fish," as the inventor had called them, were taking their uncaged fury out on the crowd. The circus tent toppled, blotting out two of the cameras. A man was tossed at a camera and left the lens shattered, the image flashing to blue before switching to another shot of carnage, and another, and another.

Deadpool Hoodie wailed in terror.

"What's happening?" the inventor shouted over the man's scream.

"The ESPs are down!" Nia called back.

Sara Jane shot the programmer a startled look. "Are we safe?"

"The walls are protected," Harrison said, a nervous waver in his voice. "Those cinderblocks are loaded up with salt. But without the ESPs, anything could come right through the front door."

"*Are we safe?*" Sara Jane repeated, the chaos on the screens reflected on her glasses.

Harrison's hand began to tremble on the mouse. He shook his head, a defeated look in his downcast eyes. His glasses slipped down the sweaty bridge of his nose and he didn't bother to push them back up.

An angry buzz came from the door. Everyone startled, turning to look. Terrified. Awaiting the inevitable. Lilian grabbed Ben's hand and squeezed it.

The buzz came twice more in rapid succession, followed by pounding on the metal door. It was the intercom. Someone was locked out there.

"Oh, dear God..." Sara Jane breathed.

"*Jesus—let me in!*" a man screamed over the intercom, trapped outside with the ghosts.

"Danny?" Harrison said, leaping to his feet.

Sara Jane put a hand on his chest to stop him from moving toward the door. "We can't let him in."

"But... it's *Danny*."

"Harrison, Danny is dead." The programmer shook his head at this. "*Yes*," she said, taking him by the shoulders. "And if someone doesn't get those ESPs back up and running soon, we're all dead too."

Harrison nodded solemnly. She let him go and he returned to his terminal.

"Is there anything we can do?" Allison asked.

Outside the door, the Ghostland employee began to scream.

"There is," Sara Jane said. "You can *pray*."

PART 2
GHOST VIRUS

The last living ghost looked upon the burning ruins of the world he'd conquered and thought, *This land belongs to us now—a world of the dead, a world full of ghosts.*

— Rex Garrote, *Shōki*

Most apparitions have very limited abilities to affect their surroundings. Others, like Revenants or Tricksters, can be very dangerous when encountered. It is advisable that Ghost Hunters avoid contact with them at all times, except under the safe conditions at Ghostland.

— *Know Your Ghosts:*
A Guide to Ghostland

DEAD RECKONING

LILIAN HAD NO intention of praying, but as she looked over the countless scenes of murder and destruction on the security monitors, she thought she might soon reconsider.

It looked like the end of the world out there. People running, screaming, trampling each other, blood everywhere she looked. A man's head exploded —actually *exploded*—and splashed the camera with his brains. A young woman ran with a little boy in her arms and a black, ghostly shape dropped out of the sky, tore her child from her arms and left her weeping. A Vietnam veteran in ragged green camouflage stalked through the chaotic crowd, mowing people down with a machine-gun. A hangman slipped his noose around a man's head and pulled, the muscles in his back rippling. The man's head popped off like a champagne cork, spraying the walls with blood.

Outside the control room door, the screams stopped abruptly.

"Danny?" the programmer said meekly.

How long until one of them gets in here? Lilian wondered. *Minutes? Seconds? How much longer do we have to live?*

Ben reached into his pocket and took out his bottle of pills. He let go of her hand to twist off the cap, shook two into his palm and popped them into his mouth. He chewed them and swallowed hard.

"What's the ETA on our ESPs, Nia?" Sara Jane said anxiously, looking back over her shoulder from where she stood in front of the monitor wall, surveying the carnage, almost in silhouette.

"My computer just restarted. Two minutes, maybe three?"

"Get it done! Harrison, have you isolated the virus?"

Sweat flicked off Harrison's forehead as he tore his gaze away from the control room door and blinked hard at his computer screen. "Nothing yet!"

"Find it and get it quarantined!" The inventor turned back to the bloodbath on the screens towering over her and shook her head. "We have to get on top of this or the press is going to slaughter us."

"The *press*?" Allison stepped into the aisle with her fists clenched, looking like she wanted to beat the inventor to a pulp. "This is negligent homicide on a *grand* scale and you're worried about your *image*?"

"This is my life's work, Miss Whatever-Your-Name-Is," the inventor said, approaching Allison down the aisle. "If we can *contain* this—"

"Look at that!" Allison stabbed a finger at the monitors. Lilian had never seen her so upset, so full of unbridled rage. She liked it. "That's not just a problem to be contained. Those people are *dying*. Hundreds of..." Her voice broke. "...fucking *hundreds* of innocent people are *dying* out there. *Children*. And you—"

"Don't you think I *know* that? We're doing everything we can—"

"I'm in!" Nia shouted.

"*How long*, Nia?"

"Hang on, hang on..."

Allison shook her head dolefully. "It doesn't matter how long. Even if you do get it under control, this park, your life's work, it's all over. This is what happens when you mess around with things you can't possibly comprehend. *This*," she stressed, pointing again at the monitors. "You took what might have been the most important scientific discovery of the twenty-first century and turned it into a thrill ride. You people should be fucking ashamed of yourselves."

The inventor opened her mouth to reply, then closed it and simply stared at her. There was nothing to say. No excuses left. It was an epic disaster and she knew it.

The light clattering of Harrison's fingers on the keyboard stopped suddenly. His head wrenched back, the cords in his neck standing and his arms flung out to his sides as if he was having a seizure. Then he launched out of his chair, spun in mid-air and slammed against the control room door so hard Lilian heard his glasses break.

In the stunned silence that followed, Ben slipped on his headset. "Oh, crap!" he whispered. He grabbed Lilian's shoulder and jerked her down to the floor with him. "*They're inside.*"

As he said this, the lights went out, plunging them into the dark.

RED EMERGENCY LIGHTS flashed over the darkened control room, on and off, clashing with the blue glow of the computers and monitors. Ben's pale white face took on their hues. Blue, red. Blue, red. Beneath the desk, Lilian's heart beat so fast she could almost sympathize with his condition.

She dared a peek over the back of the desks. Harrison lay on the floor breathing steadily, the broken frames of his glasses haphazard on his bleeding face. Shadows moved on the floor and the walls, in time with the emergency lights. On the monitors the chaos continued, soundless screams, exhibits collapsing, fires smoldering. But the control room itself had gone deathly silent, nothing but the drumming of her heart in her ears.

Where was Allison? Where was Sara Jane? The woman with the pink hair —Nia? She supposed they must all be hiding under the desks, like she and Ben were.

"Put your glasses on," Ben whispered harshly.

"They're broken."

"They still work, don't they?"

They worked, but she'd already seen enough carnage on the security monitors to last a lifetime. The tears, the screaming, the pleas for mercy. There was no escape, not for any of them, and if they were about to die, she didn't want to have to look her killers in the face. Better to not know. Just a quick shock—a stab, a slash or a snapped neck—and lights out.

Ben gripped her hand tightly. He was sweating. She felt his pulse pounding through his palm, hummingbird fast. He flinched, watching as something she couldn't see moved down the aisle. Reluctantly, she grabbed the glasses from her jacket and slipped them on.

Immediately she wished she hadn't. A luminescent black cloud hovered over the programmer on the floor, a being with seemingly no permanent shape or size: its ethereal, vaporous form contracted and expanded as what looked like a swarm of tormented faces churned within, hurtling themselves outward as if attempting to escape the others before being pulled back into its dark heart. A choir of whispers arose from within, all speaking over one another so that no one voice could be distinguished from the next, no words deciphered.

The creature left Harrison's body and floated past their row until its glow was no longer visible.

Allison's scream pierced the silence. "*No no no*—!"

Another woman's scream rose above hers. The inventor's, from the sound. And it seemed like she was running away, the scream traveling. Then it ascended, as if the woman were being lifted into the air. With a crunch of plastic, it stopped abruptly.

For a moment the only sounds came from outside the building: the distant alarm and muffled screams. A head poked around the corner of the desk, and in the dip to darkness between red and blue flashes, Lilian saw only highlights of the pale face and long black hair among the shadow. For a moment, she was sure it was a ghost.

The emergency lights came back on and she saw that it was Allison, her eyes wide and brimming with tears. Lilian let out a relieved sigh as the therapist nodded toward the door.

Ben tugged on Lilian's hand, trying to pull her to her feet. But her legs wouldn't move. She was completely immobile. Scared stiff. As she opened her mouth to tell them a muted gasp escaped her lips, like in childhood nightmares where she'd tried to call out to her mother and found she couldn't make a sound.

The inventor's scream wormed its way into Lilian's brain. She could only imagine the pain the woman was enduring. It sounded absolute. Endless. She thought that—if they ever made it out of here alive—she would dream of that scream for the rest of her life, and those dreams would frighten her awake, her heart pounding the way it was now.

Ben jerked her forward. Her legs buckled and she stumbled into an embrace. He hugged her fiercely against his scrawny chest. And just like that she felt her knees unlock. Like a magic spell. One moment stone and the next, flesh again. She pushed away from his embrace, both angry and relieved that he'd forced her into action. He gave her a brief look of embarrassment, then signaled for her to follow.

In a crouch, he followed Allison up the aisle. Lilian hurried behind them. At the door, she finally dared a look back.

Sara Jane Amblin was pinned to the monitor wall ten feet above the floor. The force of impact had cracked the screens behind her, blacking them out in the shape of a crucifix, the surrounding screens displaying images of the carnage her invention had unleashed on hundreds of innocent people. Her body thrashed against them as the anguished faces of the creature Lilian came to think of as "the Swarm" hurled itself at her, holding her in place. The look in the inventor's eyes was of pure, unimaginable terror.

And very suddenly, she stopped thrashing altogether. Her chin dropped to her chest. She was dead.

The Swarm let her fall.

The programmer groaned at Lilian's side. His nose had been mashed against his face, the eyes already blackened. She'd thought he was dead. She wished he was. When she looked back down the aisle, she saw the faces churning in the luminescent cloud look in their direction.

The door clicked open behind her.

"Run, run!" Allison cried, holding it open, ushering them through. Ben got to his feet and ran. Lilian stood shakily and followed.

"Don't leave me!" Harrison yelped, reaching feebly for them.

The Swarm spun toward the sound, hurtling toward them.

She couldn't save him. No one could.

Lilian hurried into the hall, illuminated by red emergency lighting. She slipped in the spilled coffee and inadvertently kicked Danny's severed head, nearly tripping over it as it rolled ahead of her. Allison shoved the heavy door closed behind them, muffling the programmer's terrified scream as the Swarm descended upon him.

Ben already stood halfway down the hall, urging them to hurry. She ran alongside Allison, their footfalls resounding off the cinderblock walls. Her heart hammered in her throat as they took the first corner. She was desperate to look back, but didn't dare. The Swarm was right behind them, whispering, faces churning in the cloud. Even though she couldn't hear it, she knew this. Best to just keep running, despite the burning pain in her lungs, ignoring the abomination at their heels.

As they rounded the next corner, Allison overtook her and Ben fell behind. She grabbed his hand and helped him along. He was panting, getting sluggish. His eyes looked glazed. The pills. She'd seen it before. His heart pills made him groggy. Before she'd stopped hanging out with him altogether, she used to joke they made him look like a burnout.

"Come on!" she shouted, hoping to rouse him. He blinked furiously and managed to keep pace alongside her.

Allison reached the exit and slammed her shoulder against the metal door. It didn't budge. She drew back the locks and lugged it open. Sunlight flooded into the corridor. Fresh air filled Lilian's lungs, giving her a surge of manic energy. She ran down the hall, pulling Ben along behind her and outside.

Allison closed the door, leaning against it, breathing heavily.

"We made it," Lilian gasped. "I thought for sure we were dead."

Ben doubled over, planted his hands on his knees and spat on the pavement. He rose again, his eyes going wide. His mouth dropped open. He was looking past her.

Lilian turned. What she saw made her wish they'd stayed on the other side of the door.

The street was littered with bodies, at least a dozen, possibly more. A woman lay against the curb in front of them, her face a mass of flesh and gore ground into the pavement, as if something had thrown her from a great distance. To their left, a teen with a blond rattail and a wispy mustache had been shoved into the broken plastic mouth of a trash can. His dazed blue eyes stared out at them, his upper lip curled upward in an Elvis sneer. A toddler lay beside her upturned stroller. The child's head had been torn from her body —with its *Dora the Explorer* T-shirt—and lay at the foot of a man sprawled in the middle of the street, his shirt and pants torn to shreds, the flesh beneath covered in raised slashes oozing blood. Black smoke billowed against the clear blue sky from buildings in the distance. Emergency alarms wailed. Screams rose and fell.

Lilian saw all of this in a single glance. The horror of it tattooed itself onto her retinas. She turned away only to witness more death: a man impaled through the stomach near the top of an ESP pole glistening with his blood; a teenage girl strangled to death by her own hair, green eyes bulging from her purple face; a middle-aged man with his AR headset protruding from his jugular, lying in a congealing pool of gore.

She turned to Ben, trying to focus on his face, but even in this limited view there were more sickening sights. It was no longer possible to avoid death. When she shut her eyes against it, she could still hear the tormented wails of the dying, the chaotic jangle of metal and glass, the dull thudding explosions.

We'll never get out of here alive, she thought.

Ben grabbed her by the shoulders and looked her dead in the eye. "We have to go," he said. "Before it's too late."

It was already too late. There was no way she'd bounce back from a trauma like this. She didn't need Allison's professional opinion to confirm it. A switch inside of her mind had flicked. She felt it as sharply as a broken bone or cut flesh. If they ever made it out of here alive, she would never be the same.

The *world* would never be the same.

Lilian remembered reading somewhere on the internet there were fifteen dead people to each person currently alive on earth. Even if the math was wrong, the dead far outnumbered the living. She had no idea how the Recurrence Field technology worked. Likely Garrote's virus could only affect ghosts contained inside the park, within the area covered by Sara Jane's invention and the programmer's code. She had no reason to believe otherwise. But she couldn't help wonder what would happen if that code was somehow uploaded to the internet, by satellite or WiFi or some other means. If the ghosts were somehow able to escape the Recurrence Field... and if Garrote's virus continued to replicate itself, stretching out across the globe, absorbing ghost after ghost after ghost... what would become of the world?

As if sensing her thoughts, Ben said, "The world doesn't belong to us anymore. It's theirs now. A world full of ghosts."

SURVIVAL INSTINCT

SHOCK KEPT THEM quiet, lost in their own fearful thoughts while they hurried along the promenade, skirting dead bodies and running for cover when sounds of violence erupted nearby. Allison had attempted to initiate light conversation early on, simply to alleviate the tension, but Lilian had snapped at her and she hadn't tried again.

Ben forged on ahead of the others, fighting an overwhelming urge to turn around and finish what he'd started. Rex Garrote was *here*, he was sure of it now. Alive or dead, the writer had grown more powerful than ever before, amassing a legion of ghosts with his virus. It was just like *Shōki*, one of Garrote's first novels, loosely based on a figure from Chinese folklore called Chung Kwei (although Garrote's title was taken from a similar Japanese myth), about a man who becomes the literal "king of the ghosts," able to lead the spirits of the dead as an army to take over the world. It was one of Garrote's few dystopian novels and it hadn't been very well received. Critics had called it far-fetched. Most readers had expected a continuation of the *House* series. It was his first failure both critically and commercially. In later interviews he'd refused to even talk about it.

In the book, the main character lays waste to human civilization, claiming that the living destroyed the planet and the damage could only be repaired by killing everyone. It was a dark but overly complicated narrative written in a time when America was obsessed with both Japanese culture and environmentalism. When the Shōki was the last human alive, he called himself "the last living ghost," a phrase which also opens the novel.

The more Ben thought about it the more he came to believe this must have been Garrote's plan from the very beginning. To have killed himself in the most brutal method imaginable would have left a "nuclear meltdown" for Sara Jane's Recurrence Field to pick up. But she'd claimed that her team had never seen Garrote's dead energy in that house. It struck Ben that Garrote likely hadn't wanted to be seen.

He needed to stay hidden. Nobody would suspect him, nobody could have imagined something like this could ever happen. Except Garrote. He wants to control an army of ghosts. To be the Shōki. Rule the world, just like in his book. Three hundred ghosts all controlled by the world's most sadistic horror writer. Who knows what he'll do with them if they escape—

The alarms stopped suddenly, leaving a moment of eerie quiet. From the distance Ben heard a man scream, long and wavering, as if he was dying on a runaway roller coaster. Ben hadn't realized until just then that the sounds of carnage had mostly died off. Every so often they would hear a scream or a cry for help, then a long period of silence would follow during which, aside from the bodies littering the street, he could almost imagine the entire park had long been abandoned. It had been several minutes since the last incident.

The scream came again, rising tremulously. Even from a distance, it chilled him. *Writhing,* he thought. In Rex Garrote books the dying were always *writhing*. Garrote had held a special knowledge of death most genre writers lacked. He'd watched friends die during the Vietnam War, had held dying men in his arms. Hearing that awful scream, Ben could almost picture the man writhing as he died, his limbs and facial features contorted, his fingers twisted into rigid claws.

At the end of a long, lush green hedge belonging to an exhibit called Crane Gardens[1], they took another right turn around the Starlight Arcade[2]. The sound of old school video games emanated from within, tempting Ben with their pings and blips and crunchy-sounding 8-bit explosions. Ben recognized sounds from several games: *Pac-Man*'s endless chomping, the evil laughter from the *Acolytes of Azathoth*, Zeus's command of "Rise from your grave," in *Altered Beast*.

As the arcade's noise faded away the writhing scream grew louder. He sounded much closer now.

Lilian said, "That sounds really painful."

Ben agreed with a nod. He knew if he allowed himself to speak all of his fears would come flooding out and he wouldn't be able to stop them. Since they'd fled the control center, he'd been acutely aware of Allison's scrutinizing gaze toward them and he'd made a conscious effort to appear normal. The screaming man was voicing Ben's own feelings of fear and confusion and constraint. It kept his sanity from leaping out of his head in a blubber of tears and snot.

"It feels like we've been walking for hours," Allison muttered, gasping with exhaustion.

Ben had lost his map in the escape and Lilian had suggested continuing south at every opportunity would be the best route back to the entrance. He trusted her judgment and had led them south. Her skill with directions had helped them escape hundreds of zombie invasions and monster hordes during their years of gaming together. She'd figured Demont had driven them about a third of the way back to the entrance already, meaning they would need to cover that distance twice more. They'd been forced to take several detours to avoid chaos and violence, but even still they'd only been walking for seventeen minutes according to Ben's watch. He was amazed they had lasted this long.

The street circled around a run-down three-story farmhouse. Its front porch had collapsed, a mess of broken, splintered wood blocking the entrance. Behind it an old gray windmill turned lazily in the afternoon breeze. Beside the house was an animal enclosure filled with loose hay. Like all of the other exhibits, the farm was surrounded by inoperable ESP poles. The sign out front said, *Dollop Homestead*[3]*, 1947—The Making of a Monster*.

The screams arose from the animal pen, where the screamer himself had been impaled on a pipe sticking up from the ground.

As they approached the fence Ben's vision seemed to double. Now it looked like two men had been impaled on the same pole, wearing identical blue track suits with white stripes down the limbs. Twins who dressed alike. Both had landed so it looked like they were squatting on the pipe with its jagged end rising from their screaming mouths.

Ben lowered the glasses and the screaming man disappeared, leaving just his dead identical twin speared from rectum to mouth by the rusty pipe. They weren't twins at all but the same man. The screamer was the impaled man's ghost, wailing and yes, *writhing* in desperation as he struggled to escape his own corpse.

"The blood's still dripping," Ben said. "He *just died.* He was one of us."

"That poor man," Allison said. Without another word she hopped the fence, huffing from the effort.

"Will that happen to us?" Lilian asked, as Allison made her cautious approach.

"Don't say that, Lil," Ben said, though he knew she was likely right. If they didn't get out of here soon, they would all die at the hands of the exhibits, then they too would become a part of Ghostland, a part of Garrote's ghost army. Three hundred ghosts, with more dead added by the minute. He didn't like the odds.

When she turned to him, he saw any semblance of hope had drained from her eyes. The eyelids themselves looked heavy, her gaze dulled, tired and dazed.

Post-traumatic stress, he thought. *We're all suffering from it now, every single one of us.*

"We're gonna make it out of here, okay?" he said. "The world still belongs to us, to the living. They haven't beaten us yet."

He thought but didn't say, *We don't have a chance in Hell*

Lilian turned away from him and headed for the fence. He followed. As they approached, Allison was already speaking to the dead man, calmly and assuredly.

"It's okay," she was saying. "I know this must be painful for you. Try to breathe."

Try to breathe, Ben thought. *That's a weird thing to say to a dead man.*

He wasn't even sure the ghost had heard her. The ghost just kept screaming, his gaze flitting birdlike between the three of them, like a caged and beaten animal. Ben wondered what the ghost saw when he looked at them. Were they monsters to him? Was that why the other ghosts had attacked—because they'd felt *threatened*?

"He's afraid of us," Lilian said, as if reading Ben's mind.

"We're not going to hurt you," Allison told the dead man. "We only want to help."

Cautiously, she crouched beside the man, avoiding the pool of blood widening around his feet on the straw-littered pavement. She laid a hand on his head and softly, slowly ran her fingers through his shaggy brown hair.

The ghost flinched as though he felt the sensation and he flicked his bloodshot gaze toward her. His screams became intermittent, halfhearted, like a mollified baby, until they finally stopped altogether.

"There," Allison said, smiling. "Now we can talk. My name is Allison. This is Lilian and Ben. Could you tell us your name?"

The ghost looked from Allison to Lilian to Ben and back, despair and confusion in his eyes. His mouth opened wide and he howled once more, struggling against his physical body, his semi-translucent, ectoplasmic form pulling like warm taffy. The motion left trails in the air surrounding him, the way drug hallucinations looked in movies and TV.

Allison stroked the man's hair. The ghost allowed his mouth to close, then immediately opened it again. Black blood spewed from his lips and splattered on the pavement between his sneakers, vanishing when it struck the ground. He uttered a single raspy breath—a sound Garrote might have called a "death rattle" in one of his books—and followed it with a single word: "*—dead—*" It sounded like a question.

"Yes," Allison said gravely. "I'm afraid so."

The ghost tried to look back at himself, at the pipe speared through his torso and throat. His gaze settled on her shoe, which had streaked a print in the mess of blood and straw. He seemed to note this without emotion. Then he peered around in sudden terror. "*Ghosts—high up—did you—see them?*" The echo of his voice seemed to come before the words instead of after them. It was a strange effect that sent another chill up Ben's spine.

Allison said, "Is that what did this to you?"

The ghost nodded, his head moving in slow motion, trailing faint echoes. It was disorienting and Ben found it difficult to look at him, but he didn't want to turn away. He worried the friendly ghost would turn on them if he did, the way it would have in a Rex Garrote book, becoming suddenly vicious and tearing them to pieces.

"*Running,*" the ghost said in its strange, stunted voice. "*Screaming —everywhere—flying—*"

"It's okay," Allison said soothingly, stroking his head. "It's over now. Whatever it is, it can't hurt you anymore."

Feebly, the ghost struggled against his corpse.

"Here. Take my hand." Allison raised her hand from the dead man's head and held both of them to his ghost, palms up. He looked at her hands and back to her eyes fearfully. Ben couldn't decide if the man didn't trust her or if he was afraid of what might happen now that he had a real chance of escaping his body. He thought he might feel terrified by the prospect himself.

"It's okay. You can trust me. I only want to help."

As the ghost peeled his ethereal hand away from his dead flesh, the wrists and fingers dripped like wet paint, splashing on the ground and vanishing. The ghost reached out slowly, warily toward Allison, as if he was afraid touching her might make him disappear. When their hands met his fingers slipped right through hers. Allison's eyes went wide, but not in fear. She was amazed.

"I *feel* it," she gasped.

The ghost flicked out his wrists suddenly and snatched her hands. Ben sucked in a nervous breath, certain this was the end for Allison, that the ghost would drag her to the ground and strangle her, break her bones, slit her throat, and there would be nothing they could do to help her. But the ghost merely looked at their entwined hands with astonishment.

Ben sensed an intimacy to the experience and despite feeling a twinge of embarrassment, as if he'd walked in on the two of them having sex, he found he couldn't look away. This was something no one had ever seen or would likely ever see again: a ghost leaving its body, shedding its corporeal self. A birth in reverse.

After all the death they'd witnessed in the last half hour, he was surprised to feel something close to joy, although he knew that soon the Garrote code would gobble up all the ghosts, like Pac-Man on power-ups, absorbing this man's essence as it had the others. It was only a matter of time for all of them, but this man's time was shorter.

Allison slowly got to her feet, drawing the ghost out of his body. The man moaned in pain, his limbs dripping a translucent ectoplasmic substance onto his corpse until finally he hovered about an inch above the ground, a fully freed, full-bodied phantom.

"*Thank you*," the ghost said. His voice still had that strange pre-echo quality but the ghost himself appeared more solid now, as if being free of his physical self had made him more real. Fear widened his bloodshot eyes suddenly, and the ghost turned to look behind himself. "*Be careful*," he whispered. His voice had lost its echo. "*He's coming for you.*"

"Who?" Lilian asked, looking past him, toward the collapsed porch and the boarded-up windows.

"*The writer,*" the ghost said. "*He sees through their eyes. It's so red...*" He shuddered and his face flickered, a shadow of something dark momentarily flashing over his features. Was it Garrote? Ben didn't know, but they would have to get away from him soon. They wouldn't be safe around him much longer.

"*I can feel him,*" the ghost said. "*He's trying to get inside of me.*"

"Can you fight him off?" Lilian asked.

The ghost nodded. "*I don't know for how much longer. Thank you for freeing me. I have to go before—*" He scowled, looking down at his own hands. He seemed confused by them, as if they belonged to someone else. "*I don't want to hurt you,*" he said.

Allison gestured with a nod. "Go. Stay safe."

The ghost smiled and thanked her again. He turned and hovered toward the far edge of the fence, overjoyed to have found his freedom, and Ben wondered what it might feel like: to escape the burden of his body, to never have to worry about overexerting himself, to travel anywhere and do just about anything he wanted, to live without the specter of death constantly hanging over his head. He thought it would be incredibly freeing, so long as he wasn't trapped within Ghostland's walls.

"'He sees through their eyes,'" Lilian repeated with a melodramatic shiver. "Well, I'll never sleep again."

"Wasn't that *thrilling*, though?" Allison held out her hands to them, marveling at something Ben couldn't see. "I can still feel his touch on my skin."

Ben thought to tell her not to get the pottery wheel ready just yet, but the ghost's warning echoed darkly through his mind. *Did Garrote get inside of me that day? Is that why I feel so connected to that house, why I felt such a strong urge to come here today? Is Garrote seeing through my eyes, too? Manipulating me like he's been manipulating his ghosts?*

If he did die today, Ben wondered how much of his memories, his *essence*, would remain when the Shōki inevitably absorbed him into his army of ghosts.

CO-OP MODE

"THERE'S ONE!"

Out front of the windmill, the newly formed ghost fled two large men in black and white uniforms. Ben recognized them at once: Niko, the security guard who'd told him he looked like he had a strong heart, and his partner, Leonard. The AR headsets jostled on their faces as they chased the frightened ghost.

Niko raised his left hand and aimed what from a distance looked like a remote control. Blue and white sparks discharged from the end of it and Ben realized Niko was holding a stun gun.

"No!" Allison cried out.

The charge struck the ghost in the shoulder and for a moment he became brighter, a being made of pure light. Then each one of his billions of molecules exploded outward—Ben thought of a dying star—and vanished in wisps of thin multicolored smoke.

As the two guards gave each other a high five and laughed, Allison hopped the fence and bounded up to them.

"*Why the hell did you do that?*" she shouted, getting right up into Niko's face, standing on her tiptoes with her hands balled into tight fists. He was at least a foot taller than her and outweighed her by more than hundred pounds, but righteous indignation had overtaken her common sense. She was mad as hell and she wanted an explanation.

The two men looked at each other as if Allison had lost her mind. "What do you mean?" Niko said. "It was a ghost."

"He was *one of us*!" Allison shouted. "He wasn't going to hurt anyone!"

Leonard, with his bald head and goatee, snorted laughter. "Is she for real? Lady, we been attacked by half a dozen of those things. How are we supposed to know which ones are friendlies and which ones want to take our damned heads off?"

"Fallujah," Niko said to his partner, and Leonard nodded. Niko, the man Ben thought looked Hawaiian or Maori with his brown skin and tribal tattoos, looked over Allison's shoulder as Ben and Lilian approached the group. "*Heyyy*, little dude! How's your ticker?"

"I'm okay. How did you know that would work?"

"With the stun gun? It was just instinct, man. Those things scare the heck out of me, big time."

"You're afraid of ghosts and you took a job at Ghostland?" Lilian asked in a condescending tone.

Niko shrugged. "It's a good-paying gig. Never figured they'd get out of their cages, did we, Leonard?"

Leonard nodded. "This was definitely not in the training manual."

"You really shouldn't have done that," Allison muttered.

"Look, lady—" Leonard began.

"My name is Allison!"

"Great. I'm Leonard." He wiped his right hand against his chest and held it out for her to shake. After a moment he retracted it, unshaken. "Good to meet you," he said anyway. "This is Niko. Now that we're all acquainted, let's see if we can come to an understanding. I'm sure we've all seen these things turn on people. One second, they're doing their thing and the next they're tearing someone's throat out. We don't exactly have the time to second guess."

"You *chased* him," Allison said bitterly. "He was running away."

"Technically he was floating, ma'am," Niko said.

She flashed him a look so cold that he cringed and took a step back, raising his hands. "I, uh... I forgot your name," he said nervously.

"*Allison*," she replied through gritted teeth.

"Allison, right. Sorry. And sorry about the, uh... your ghost friend there. We kinda just got caught up in the moment, didn't we, Leonard?"

His partner shrugged. "I just think it's pretty cool the way they—" He looked at Allison, who was staring daggers at him, daring him to continue. "I mean they kind of look like fireworks," he finished meekly.

"Seriously, though," Niko said, scowling at his partner. "We didn't know. I just figured they were all... you know. Because the others tried to kill us. If I could take it back, I honestly would." At this he pressed the hand holding the stun gun against his heart.

"What's done is done," Allison said, looking down at her hands as she rubbed them together methodically, as though she could still feel the ghost on her fingers. "Dwelling on it isn't going to help."

"I agree," Leonard said. "Me and Niko were talking about that earlier, weren't we, Niko?"

"That's right."

"Tell them what you said to me," Leonard prompted.

Niko gave his friend a look Ben couldn't decipher. "I was saying, we can sit around freaking out, waiting for someone to come save us, or we can fight our way out of this place."

"That's what we're doing," Ben said. "But why didn't you guys just walk out the front gate?"

The guards shared another indecipherable look.

"Entrance is closed up tight, little dude," Niko said. "Security protocol. Big metal gate came crashing down, landed right on this dude's head, crushed it like a friggin' watermelon. Three feet thick of solid impenetrable steel. There's no way in or out of this park until the protocol is lifted by one of the big bosses."

"Except for the service hatch," Leonard reminded him.

"Right. That's where we're heading now."

Allison looked up from her hands. "Service hatch?"

Niko nodded. "In the basement of Garrote's house. It's supposed to lead out to the mechanical building on the other side of the fence, only..." He trailed off, looking askance at his partner. Leonard looked down at his own boots.

"Only what?" Lilian asked.

"Well, we don't have the door code," Leonard admitted.

"But a mechanic buddy of ours does," Niko added hopefully. "We just gotta find him."

"And hope to God he's still alive."

"Right." Niko sighed, no longer sounding hopeful.

Ben tried to put their plan together in his mind. It sounded like a longshot, a hail Mary that would only work if all the pieces were in the box and fit together perfectly. "So, you're saying the only way to get out of Ghostland is through a service hatch in the house of the guy who's trying to kill everyone... and we don't even know if we can open the hatch once we get there?"

Niko and Leonard shared another look.

"I mean, basically, yeah," Leonard said.

Allison groaned. "Terrific. Do you happen to have any idea where this friend of yours might be at this moment?"

Niko and Leonard both made to answer. Leonard nodded for is partner to speak.

"Last we heard over the coms he got sent out on a service call," Niko said. "Something about the automated doors malfunctioning in the asylum."

"Asylum?" Lilian said, toying with the beads of her totem, her magic charm. Ben put a hand on her shoulder. She flinched away from him.

"With all the places he could go," he said, "I don't think Morton Welles would be sticking around there."

"I guess we'll have to risk it. No other way out, right? Not like we can walk out the front door or anything. No. The animals are out of their cages and they've got a stupid *protocol*."

"Hey, don't you worry, pretty lady," Leonard said, cracking his knuckles. "You've got two big strong men here to protect you now."

"Three," Ben said, his cheeks flushing with resentment.

"She's *seventeen*," Allison said sternly.

"*Eighteen*," Lilian corrected her.

"Hey, hey," Leonard said, raising his hands in mock surrender. "Let's not jump to conclusions here. I was just being nice."

Allison eyed him with suspicion. "Yeah, well be nice from a respectful distance."

The big man took two paces back and grinned smugly.

Niko eyed his partner. "Ma'am—*Allison*—we only want to get out of here, same as you all. We're all geared-up. You're better off sticking with us." He gave her a toothy smile. "So let's team up, huh?"

They all agreed. Even Allison nodded begrudgingly.

Ben wasn't sure how much they could trust Leonard, but Niko was right. Not only was there power in numbers but they could stand to have someone with weapons looking out for them. If only there were enough stun guns for the rest of them, he would have felt a lot safer.

Especially now that they were going back, straight into the belly of the beast. Garrote wouldn't let them walk right into his house without a fight. It wasn't going to be easy. But five lives were better than three. And even if Leonard and Niko eventually turned out to be redshirts, at the very least they might provide cover for the rest of them to escape.

Ben didn't like to think in those terms but that was how things were now. The best he could hope for was for Lilian and himself to make it out alive.

He didn't want to consider the worst.

"ALL RIGHT," Niko said, hiking up his pants by the belt. "Let's get moving, huh? Maybe we can get out of this damn place before nightfall." The big man took a step toward the east.

"The asylum is that way," Lilian said, pointing back the way they'd come.

Ben nodded. "She's right."

Niko squinted off in the direction he'd been heading. "Sure, of course it is. Why don't you lead the way? It's Lilian, right?"

She started around the farmhouse fence. "That's right. And don't forget it."

Niko chuckled, following along. "I gotta say, you three did a damn good job staying alive so far. You're the first living people we've seen since about ten minutes into the apocalypse."

"And we didn't even have any weapons," Ben said, hoping to make his point without belaboring the fact that he was just as capable as the two security guards.

Niko nodded. "That's pretty impressive."

While they walked, he explained that he and Leonard had been stationed in Afghanistan—what he called the "Sandbox"—together before their battalion had been pulled out. Back in the States, the two of them left the Marines and drifted around taking any odd job they could find. By the time they saw the postings for Ghostland they'd already drifted apart and been living in different states for little over a year. Niko had called Leonard at Leonard's parents' house to tell him about the job. In a strange coincidence,

Leonard had just circled the ad himself when the phone rang.

"It was kismet," Niko said.

"Or so we thought."

Niko had taken the bus out from Florida to meet Leonard in Tennessee and the two of them had driven to Duck Falls together in Leonard's old family truck. They'd scheduled their interviews on the same day, a few hours apart. Niko had waited in the truck during Leonard's interview, and vice versa. Leonard had been sure he'd botched it. Niko had felt hopeful. They'd stayed overnight at a bed and breakfast in town (Ben knew the owners through his mom—"Nice folks," Leonard said), and both heard back the next morning within minutes of each other.

"After all that time drifting, we were happy as pigs in shit," Leonard said.

Niko nodded. "And it was a pretty sweet gig for a hot minute. Decent hours, no night shifts, overtime pay. We rented a killer pad in downtown Duck Falls—"

"That's where I live," Lilian said. "I've never seen you guys around."

Leonard shrugged. "Truck's usually parked out back of Green's Antiques. We do most of our shopping in Hagerstown and we're on rotating shifts."

She said, "I live right across the street from you."

"Wow. Isn't that weird?"

Ben noticed Allison watching the exchange with suspicion. Leonard caught her watching him and cleared his throat before busying himself by fiddling with his stun gun.

The non-disclosure agreement had nearly gotten them into trouble, Niko told them. Everyone employed by Ghostland had been required to sign it, to declare they wouldn't reveal any "proprietary information" about what they saw inside or face legal action.

"Man, I wanted to blab so bad," Leonard said. "Can't count the times I came close on a good bender at The Blind Duck—"

"I hate that place," Allison said. "Always so rowdy out front, especially at night."

"Well, me and Niko have been known to get a little rowdy whilst in our cups, hey, bud?"

Niko chuckled like a man with a secret and the two of them continued their story. During one of these rowdy nights, Leonard had come closest to spilling the beans. They'd been "putting the moves," in Leonard's words, on a couple of local women when one of their colleagues had swooped in, asking the women if they had ever seen a ghost.

"Just as this dude's making his pitch to drive the ladies out and show them this place, our supervisor walks up, out of nowhere," Leonard said. "I didn't even know she was in the bar."

"In our defense, we were pretty hammered," Niko added.

"And laser-focused," Leonard said, causing Allison to roll her eyes at the implication. "Anyways, this dude gets fired right there in front of us. I didn't feel like blabbing so much after that."

"I think I read about that in the *Squawker*," Ben said. The men looked confused so he explained. "It's the town paper. The Duck Falls *Squawker*."

"That's a pretty good name, actually," Niko said. Leonard agreed with a shrug and a nod.

Once Ghostland had made the announcement the job had felt much more secure, Leonard told them. And Niko had never been required to deal with any of the ghosts personally. "They got these lab techs with silver spacesuits, if any of the ghosts start acting weird one of them gets sent in to take care of it," he said.

"I saw one," Ben said. "In the hangman exhibit. How do those suits work?"

Niko shrugged. "Beats me. Looks like one of Leonard's clubbing shirts when he forgets to add the fabric softener."

Leonard chuckled.

Talk of the lab techs brought Niko around to opening day and the chaos they'd seen before meeting up with Ben and the others. They'd watched people grabbed by flying ghosts—"Reminded me of those flying monkeys in the Wizard of Oz," Leonard said—and dropped to the ground, people torn in half, people riddled with phantom bullets and slashed by knives and fingernails, hair pulled out, eyes gouged, intestines ripped from abdomens, skulls stomped, bodies set on fire—

"*Ugh*, we get it," Lilian groaned.

"My bad," Niko said. "I guess me and Leonard are kinda desensitized to it, after what we went through in the Sandbox. Anyways, we tried to help as many people as we could but once we realized it was pretty much a lost cause we got geared up and moving, didn't we, Leonard?"

"*Oorah!*" Leonard grunted.

Ben suddenly remembered the exchange the guards had at the gates with the guy wearing the No Fear T-shirt. "Did you guys bring that gun?" he asked. Niko and Leonard just looked at each other, oblivious. "The handgun," Ben prodded. "The one you confiscated."

"Oh, the *peashooter*!" Leonard said. "Nah, we didn't bring that. Forgot all about it, now you mention it. Not that it would do us any good against these things anyway, right?"

"Yeah, I guess so," Ben said, trying not to sulk. In a Rex Garrote book, if a gun was introduced in the beginning it would show up again at the end. He worried that sooner or later they would find themselves in a situation where a gun would be necessary, and they'd kick themselves for not bringing it along.

"So where were you three when the shit went down?" Niko asked after a moment of dispirited silence.

As they passed by the arcade going in the opposite direction, Ben told them about Demont, about the control center and Sara Jane Amblin and Rex Garrote's ghost virus. He told them about what had happened to the inventor and the programmer, and both men seemed genuinely dismayed.

"So everyone in charge is dead," Niko said. "Anyone who could stop this, get those front gates open."

Lilian said, "Demont might still be alive."

Ben agreed that might be so, though he doubted it. Demont had been on his own when he'd left them. The chances of him having survived were pretty low.

"What I don't understand is why isn't anyone coming to rescue us?" Allison said. "If there's a security protocol, shouldn't the first response be to call the authorities?"

"Landlines are down," Niko said. "And you mighta noticed you aren't getting any bars on your cell phones—that's because the park tech obstructs cell networks." Lilian nodded at this as if she had already noticed, and Niko continued. "The security system is supposed to autodial to the police during a shutdown but with the landlines down we're pretty much S.O.L."

"What's S.O.L.?" Lilian asked.

"Shit Outta Luck," Niko said.

Lilian's shoulders slumped.

"We're on our own here," Leonard said bitterly, watching his boots scuff the pavement. "Nobody's gonna save us but ourselves. The sooner we all get used to that the better."

"You're right," Lilian said.

Leonard looked up at her and she gave him a brief sad smile. He resumed looking at his boots, grinning to himself. Ben watched them and wondered if Allison hadn't been right to accuse the man of getting too friendly. Something about that shy, boyish grin troubled him.

It must have troubled Allison too. She was eyeing the man in distrust and glanced askance to raise an eyebrow at Ben.

ASYLUM

EVENTUALLY THE SURVIVORS found themselves back on the now-deserted promenade, where Lilian pointed out the strange creatures congregating there. "Look! Fireflies!" she said.

Ben had seen them a moment earlier and dismissed them as dust motes caught in the sunshine, a trick of light through the glasses. Looking closer, he realized they looked more like the embers of a nearby fire.

"What are they?" Allison said, watching the colorful things pitch and yaw. There were dozens of them spanning the width of the promenade, maybe hundreds. Most were light blue, orange or peach but there were various other colors as well.

"I think they're orbs," Ben said.

Allison scowled. "Orbs? I thought those were hoaxes?"

"Well, yeah, usually." Ben watched the strange creatures meet in a swirling mass at the center of the promenade, like a bright, colorful swarm of gnats. "But these are real. They're actually real."

Lilian reached out to touch one.

"Careful," Allison said.

Lilian ignored her. The orb spun around her outstretched hand, leaving a fiery orange trail like the end of a sparkler and making her cupped palm glow for a moment, before it fluttered off to meet the others.

A light blue orb floated past Ben's face and fluttered down toward the concrete. It landed on a crumpled Milky Way wrapper, which began flattening itself out with a crinkly sound. The flattened wrapper rose an inch or two into the air and sailed for about a foot like a leaf on a breeze, then dropped back to the ground. It glowed bright blue for a moment before the orb withdrew from it and caught up with the rest.

"They're almost like poltergeists," Ben thought aloud.

"Imagine what more than one of them could do," Allison muttered.

"I think they're kinda cute," Leonard said. The others looked at him, surprised by the sentiment. "You know, for ghosts," he explained with a shrug, and everyone laughed. It felt good to relieve the tension. But Ben knew their relief was temporary. Death was waiting for them just around the corner. He was certain they would have their fill of it by the time this day was through.

THEY CONTINUED their game of duck and cover heading north, leaving the orbs dancing behind them. Within a few minutes Ben and the others stood beneath the asylum, looking up at its brick towers, its parapets, and tall, darkened windows behind black iron bars.

"Entrance is around the other side," Niko said, pointing off to the right.

They followed him around the corner of the building, through an arched walkway and up a set of wide steps. The cemented seams were visible where the building had been cut into portable pieces and reassembled, fitted with steel rebar for support. The entire process to disassemble, transport and reassemble these buildings must have cost a fortune. Garrote's estate had been worth several million when he died, but it wouldn't have been nearly enough to bring a place like this to life. Ben knew very little about Ghostland's partner company, the Hedgewood Foundation, aside from the fact that they were a Fortune 500 company who seemed mostly interested in industrial software and green technologies.

Large columns stood on either side of the wide, arched double doors, the words BRIGHT FALLS SANITARIUM etched into the stone above. A golf cart had crashed and tipped partway up the steps. The driver lay sprawled beneath it. Lilian didn't seem concerned with the dead body. She was looking up at the doors, and when she looked back at Ben her eyes were wide with fear.

"It's gonna be okay," he said. "If he's in there... we'll hit him with a blast of electroshock therapy. Light him up like the Fourth of July."

"I don't know, Ben. I feel like we're about to walk into a bloodbath. I mean, we don't even have a plan..."

"Like a young Leeroy Jenkins," Ben said, forcing a grin.

"Yeah," she said, unable to keep the dread from her voice. "Only everybody on his campaign died. Including Leeroy."

"Let's get this over with," Allison said, slipping past them and ascending the steps. She grasped the handle and pushed on the door to the right. The hinges groaned as it disappeared into the darkness within and Allison looked back with a nervous chuckle. "I guess nobody's home."

Niko stopped alongside her. "Probably best if I take the lead," he said, holding up his stun gun and waving it. "Leonard, watch our six."

"Roger that," Leonard said. He saluted Niko with a childish giggle that made Ben wonder if he was stoned. But he was probably just scared, like the rest of them.

Niko stepped into the darkened interior. Allison followed close behind. Ben and Lilian entered side by side. When Ben turned back, expecting the door to slam shut on them, he saw Leonard scanning the courtyard briefly before stepping in behind them.

"All clear, my man," the ex-Marine said, pushing the door closed.

The foyer was wide and tall, a full three stories, and several degrees cooler than the outside. The walls and arched ceiling were dark-stained wood, the floors checkered black and white. A circular administration desk stood in front of the doors. Beyond it lay narrow tracks and an operator platform for a train the size of a roller coaster. The tracks led down the long, arched corridor to the left, where sunlight filtered into the foyer. On either side of the foyer stairwells rose to a second story. The tracks returned down a low slope alongside the staircase to the right, closing the loop.

"I did my residency in a place like this," Allison said as she headed cautiously around the desk, fingering its surface. "Back when these places were still common. It wasn't what I expected, not like in the movies. It was... quiet. Everyone heavily medicated. The night shifts were a little nerve-wracking with the whole hospital practically to myself, only a few others on duty. One night, I was consoling one of the patients after he'd had a particularly bad night terror. He was delusional but he'd never been violent. Always very meek."

She gave Ben a sad smile. He wasn't sure where her story was going but she seemed to need to get it out, likely because of their brush with death at Guest Services.

"Somehow," she said, "he'd gotten hold of a sharp piece of metal—a hinge of some sort, I believe. Before I knew it, he'd slashed both of his wrists. Blood everywhere, on the bedspread, on my clothes. I tried to stem the bleeding, I called for help but the building was just so *large*. So empty, just like this place. I had to leave him there bleeding to run and get medical supplies, and by the time I'd returned with the overnight nurse who'd been on her smoke break he'd already bled to death. It was the first time I'd seen someone die. One of my patients, I mean."

Lilian looked at her curiously. She opened her mouth to speak, maybe to comfort her, but a long groan of wood from the second floor interrupted her, followed by the hurried step of hard soles on tile, approaching from somewhere deep within the asylum.

"Maybe it's our mechanic friend," Allison suggested cautiously.

"Probably shouldn't wait around to find out," Niko grunted. "I don't know about you guys, but I've seen enough ghosts for one day."

"That sounded like lady steps to me," Leonard said. Allison flashed him a look of vague annoyance. "But he's in here somewhere," he added. Then he nodded as if to confirm this to himself. "Yeah, he's in here. Let's head down that hallway, huh?"

Ben peered into the small operator platform as they stepped over the tracks. A man in a red Ghostland T-shirt was slumped over the controls, blood crusted beneath his eyes and open mouth. Something had sliced his forehead cleanly from temple to temple and peeled back his scalp and hair, revealing the skull beneath, a trench carved out of the bone as if someone had used a very sharp tool, possibly even a bone saw.

Beside the operator platform was a bin filled with VR controllers. Ben grabbed one and flicked it on, and the plastic bulb on top glowed red. It left trails on his glasses as he swished it back and forth. There had likely been a mini-game on the roller coaster where the controllers could interact with ghosts. He wished he'd been able to enjoy it before all hell had broken loose.

Leonard and the others were already following the tracks to the left. Ben hurried to catch up, tucking the controller into his backpack. He met them at a glassed-in corridor through the middle of an outdoor garden. Small neatly trimmed hedges stood around a large apple tree and rows of multicolored flowers, and carved stone benches were placed strategically in front of birdbaths, fountains and statuary. They could see both wings of the asylum on either side of the quadrangle. It just looked like a whole lot of doors on either side, probably patient rooms.

As they headed down the corridor, the vines draped over the tunnel made light scrabbly sounds like fingernails against the glass. The apple tree swayed in the same light breeze. Several apples fell from its branches with soft thuds.

Ben was watching the gnarled branches sway when a pale woman wearing a nun's habit stepped out from behind the tree, stooping to pick an apple from the ground. She was naked from the shoulders down, her nipples as red as the apple and pubic hair as dark as the tree. She brought the apple to her nose, her forearm pressing her breasts together momentarily, and she inhaled deeply. As she bit into the apple, she looked up suddenly and stared directly at Ben. He felt his shorts getting tighter, his penis beginning to throb with the beat of his heart. He swallowed hard.

"*Do I tempt ye?*"

The voice, with its Scottish lilt, seemed to come not from her mouth—chewing slowly as she spoke—but from inside Ben's head. He looked at the others. They hadn't noticed her. The nun's crazed eyes followed him as he hurried ahead to catch up.

"Did you guys hear that?"

"I think it's just the vines," Lilian said.

Ben peered back through the cracked glass. The nun was gone. The branches of the apple tree swayed in her absence.

They passed through an archway into a dim corridor with battered lime-green metal doors set in its stone walls. Their footfalls echoed hollowly as they moved past a rusted wheelchair tipped over in the corner, a wheel spinning with an incessant rusty squeak. As Ben glanced back into the garden, Leonard eyed him with curiosity.

"Something on your mind, bud?"

"You didn't see a woman back there, did you?"

"What woman?"

"What woman?" Allison echoed.

"I thought I saw a nun in the garden," he muttered. "Eating an apple."

They all paused to look at him. Allison had raised an eyebrow. "Well, that's rather suggestive."

Ben blushed.

Niko said, "Hey, what's that?"

The group's attention shifted to the security guard, and Ben breathed a sigh of relief. They'd been looking at him like he'd experienced some sort of perverted hallucination. After everything they'd seen it hardly seemed fair to be skeptical.

Niko crouched to look at a dark heap on the grimy green linoleum, tucking the stun gun into the holster on his belt. He picked up the pair of dark blue coveralls, and gave Leonard a worried look. "These are Joe's."

"Dang," Leonard said.

"At least that means he's here somewhere," Niko said, folding up the coveralls and tucking them under an arm.

"Runnin' around butt-ass naked?"

The survivors shared a wary look before continuing down the hall.

"It's cold in here," Lilian said. She was shivering, hugging her arms to her chest.

"It is," Ben agreed.

"I run hot," Leonard said with a shrug, and Niko laughed.

"Look there. Those must be his boots." Allison pointed to the end of the corridor where it turned to the right, where a pair of boots lay haphazardly in the light of a dirty window, as if the mechanic had kicked them off in a hurry.

"Are those Timberlands?" Leonard asked. "Size twelve?"

"You're not stealing Joe's boots, Leonard."

"I wasn't going to take 'em." He turned to Ben, giving him an innocent shrug. "Just making sure they were his."

Ben wasn't sure he believed the man but he didn't care. All that mattered was finding the man who owned them and getting the hell out of this place. They had to reach the mechanic before it was too late, if it wasn't already. They needed that security code.

"Ah, Christ, here's his clipboard," Niko said, stooping to pick it up.

"You sure that's his?" Leonard asked.

"Who's else's would it be?"

"I dunno. Don't they use clipboards in hospitals?"

"This ain't a hospital anymore, Len. It's an exhibit."

"Right. Well, maybe it's like a prop."

"'Fix mechanical door wiring in asylum,'" Niko read. "'Crane Gardens fountain requires adjustment. *Angel Knives* machine eating quarters again.' Still think it's a prop?"

"Guess not."

Lilian shivered. "You guys, it's *freezing* in here. Like, seriously. I can see my breath." She demonstrated, exhaling sharply. Ben did the same. Their breath plumed out in front of them.

While they marveled at this, like kids out making snow angels, a door at the far end of the long hall swung open and slammed against the wall. The sound echoed throughout the asylum—*BANG-bang-bang-bang*.

The five of them stopped in their tracks. Listening. Waiting.

With a high-pitched shriek, a hairy man in tube socks and tighty-whities staggered out of the room, visibly trembling. White flakes fluttered out behind him, caught in the light from the garden windows.

Lilian blinked rapidly, as if she couldn't believe her eyes. "Is that... *snow*?"

"Joe?" Niko called out. His voice resounded off the cold stone walls. He made no move to help his friend, looking nervous as he drew the stun gun. "Are you all right, brother?"

The man at the end of the hall, hugging himself and shuddering in his boxer shorts, turned at the sound of Niko's voice. He opened his mouth to speak, managed only a slight whimper before his legs seemed to be jerked out from under him, and he struck the floor face-first with a slap of flesh and a cry of pain. He rolled onto his side, staring back into the room. "N-no," he said, shaking his head in absolute terror. "*No, please—!*"

The unseen force yanked him back into the room, his flesh squeaking on the linoleum floor. For a moment his screams of agony echoed down the empty corridor. Then the door slammed shut behind him, muffling them.

A woman's high, throaty laugh, taunting and malevolent, pealed out.

Niko broke into a full run, calling out his friend's name. Reluctantly, hiking up his pants by the belt, Leonard chased after him.

Ben turned to the others. Lilian seemed as frightened as he felt, twisting her beaded bracelet around and around her wrist and moving her lips. She caught his look and shook her head, a small, almost imperceptible movement that barely moved a strand of hair. "It was snowing in there," she said. Allison was looking behind them, back down the hall the way they had come, as if she'd seen or heard something the rest of them hadn't.

Ben followed Niko and Leonard, heading down the middle of the hall, wary of the windows to his right and the doors to his left. They were closed, but he fully expected any of them to burst open at any moment. He peered into the garden, certain the nun would be standing under the apple tree, juice spilled down her chin and running down her breasts. But the quadrangle was empty. The apple tree swayed gently and a fine mist from the fountain, with its green marble cherub pouring water from a jug into the murky pool below, swirled in the breeze.

A hand fell on his shoulder, startling him. He whirled around to find Lilian standing beside him with a look of worry. "Did you see something?"

He shook his head. "No. It's nothing."

Ben and Lilian approached Niko and Leonard. The two former Marines still stood by the room their friend had been pulled into, staring at the door, seemingly afraid of what they might encounter within.

"I don't hear anything," Niko said. "You think he's dead?"

"We gotta go in there, man," Leonard muttered. He didn't sound confident. "I know you're scared but he could still be alive."

Niko nodded, his Adam's apple bobbing. "I can't open it, brother. You gotta do it."

Leonard huffed and reached out to grasp the handle. The second his fingers curled around it he withdrew the hand with a wince. The metal had left an angry red print on his palm. "Jesus, it's ice cold," he said, clenching and unclenching his fingers in a vain attempt to soothe the pain.

"Ouch," Allison said.

Tucking his left hand into the hem of his T-shirt he tried again, twisting the handle quickly and jerking the door open. He stepped out of its arc as it swung out violently and struck the wall.

More white flakes eddied out into the hall. At first, Ben thought it must be ash or pillow stuffing, until one of them landed on his forearm and melted, leaving a cold wet spot on his skin. Lilian was right. There was a blizzard inside the small room.

"Joe? You in there, brother?"

Fear in Niko's voice. Fear in his eyes. The hand holding the stun gun quivered, just enough to be noticeable, rattling the plastic parts. No reply came from within, just the wavering howl of a phantom wind as the giant snowflakes whipped in a frenzy, blotting out whatever lay inside.

"*Do ye fancy me?*"

Ben spun on his heels at the sound of her voice. *The nun*. She'd followed them. Followed *him*.

"*Aye, yer a shy one, aren't ye?*"

She uttered another high, throaty laugh, the same laughter he'd heard when the door slammed behind Joe. Ben looked around wildly, seeking its source. Was he the only one hearing her? Lilian looked at him apprehensively. He didn't know what to tell her. Whatever was happening to the man inside the room was more urgent. His life hung in the balance. So Ben said nothing, just shook his head and looked out the windows through the corner of his eye.

Inside the room the blizzard had thinned out. He could make out a cot and a small shelf, the walls painted salmon, flaked in places, cracked elsewhere, with huge chunks of plaster missing. A few drawings had been tacked to the walls, crude sketches of gaping vaginas and giant penises dripping great gobs of semen had been etched into the plaster.

The mechanic lay against the wall in his underwear, slumped over his hairy gut, his legs spread wide. A trail of blood led from his groin all the way to the door, the crotch of his boxers stained black with it. Ben shuddered to imagine the sort of injury that would have caused so much blood, especially where it was.

"Is he dead?" Lilian asked.

"Ah, Christ, I'm going in." Leonard stepped into the room. He followed along the path of blood as snow swirled lazily around his head. "Hey, Joe, buddy. You alive? C'mon, wake up, man."

Niko stepped in behind him in a quick movement that seemed as if he'd had to force himself forward, like someone wearing a shock collar trying to cross an invisible fence without getting zapped. "Is he okay?"

"It's not him," Leonard said, looking back with alarm.

The nun's laughter rattled in Ben's mind. She stood in the garden just beyond the windows, still naked from the cowl down. Black varicose veins radiated from her wide, delirious eyes, tracing their way below the cowl and along the slope of her breasts. His gaze fell upon them, hanging like withered fruit below the cowl, marked with bruises and bite marks. Her cracked and bleeding lips parted in a maniacal smile full of rotted teeth and mushy apple.

"*I've come for you, my shy boy,*" she said.

Despite his fear, despite his repulsion, the sound of her voice prickled the hairs on the back of his neck and he felt himself stiffening again. He willed it not to happen, but like his flushing cheeks it was impossible to control once it had started.

He turned to the others. They were too concerned with what was going on inside the room. No one saw the nun walk right through the glass and stride toward him, her pale flesh covered in red slashes, her dirty-nailed toes floating inches above the linoleum.

As much as he wanted to, Ben couldn't look away. He didn't dare.

She pressed her cold, fish-belly lips against his with the sickly-sweet smell of death on her breathy moan.

LILIAN HEARD the crunch of footsteps in the snow a moment before blood-red footprints appeared against the white, approaching Niko where he stood by the door. They were small footprints, a young girl's or possibly a small woman. The ghost had the power to control weather and had cut the junk off the guy on the floor. If Niko and Leonard were smart, they would hustle out of there before she decided to mess with them.

"Uh... guys?"

"What do you mean it's not him?"

"I mean it's not him. It's someone else."

"Well, who is it then?"

"How the hell should I know?"

"Guys!"

Niko and Leonard stopped bickering and spotted the footprints. Leonard backed away from the dead man on the floor while Niko drew his stun gun.

Crunch. Crunch. The footsteps came closer.

Lilian stepped back to give Niko room as he got into a wide stance, aiming the stun gun with both hands.

He pulled the trigger. He frowned and pressed it again.

Nothing.

He twisted it around to look at the power indicator on the side. "Shit," he muttered. "Out of juice."

"I told you it needed eight hours to charge," Leonard said.

A whimper came from behind her. Ben was shaking, staring into the empty space in front of him. She called his name but he didn't react. His lower lip pooched out and stretched as if something had pulled at it and then snapped back against his teeth with a slight *plip*.

She shook him by the shoulder. "Ben!"

He startled, rousing out of his stupor. It was like he'd been sleepwalking. He turned to her, his cheeks flaring red. His shamefaced glance down at his crotch drew her attention to the tent in his pants. "W-what? What's happening?" he stuttered.

"Come on," she said, consciously ignoring his erection. "We have to leave."

Inside the room, Niko said, "Leonard, we got it cornered, brother. Stun it!"

Leonard had pressed himself flat against the wall, gripping it with both hands flat. "I... can't... *move*."

"What d'you mean you can't move?"

"It's *got* me, man." He shot them a terrified look. "Something's got me!"

Niko looked down at the footsteps on either side of the door, as if the ghost was preventing him from entering. "Ah, hell," he said, and stepped into the room.

An invisible force launched him back so violently he knocked Allison to her knees and was thrown against the far wall, cracking the glass with the back of his head.

Lilian uttered a frightened yelp as the door swung shut and crashed against the jam. The lock clicked. Leonard's muffled cries echoed within.

Niko rose from his knees, pulling up his sleeves with rage in his eyes. He rushed the door and slammed his shoulder into it. "Hang on, brother! I'm comin' for ya!" He shouldered the door once more before stepping back and examining the handle. "Damn thing swings out, doesn't it?" He grabbed it and pulled. It wouldn't move. "Come on, guys, help me pull."

Leonard's screams grew louder behind the door, frantic. A howl of female rage rose to match it. The bedsprings rattled and twanged. Something thudded heavily against the door.

Silence followed.

"Len, don't you die on me!"

Lilian grabbed Niko's flexing shoulder muscle and Allison took his other arm, and the three of them pulled, their muscles straining. "Ben!" she called over her shoulder. "We need your help!"

The lock clicked open and the door came swinging outward. The three of them stumbled back as it slammed against the wall.

Leonard stood in the middle of the room, his back to the door, looking down at the dead man on the floor.

"You all right, brother?" Niko said.

Leonard's head rose. He spoke without turning. "I'm fine, my good friend. This man is dead. We can do nothing more for him."

Niko frowned. Something was off about Leonard. It sounded like he was doing an accent.

"All right, cool," Niko said. "Let's blow this joint then, huh?"

"Yes, let's."

Leonard turned rigidly on the spot. He blinked several times, like a man who'd just removed his glasses and couldn't quite focus on what he was seeing. Then he crossed to the door. Blood oozed from his nostrils and dropped onto his nametag. He wiped his nose with the back of his right hand and studied the red smear on his finger. "Do any of you happen to have a handkerchief? I appear to be bleeding."

Allison watched him cautiously as she handed over a folded tissue from her jacket pocket. He studied it a moment before pressing it to his nostril. "Thank you," he said, his voice nasal from the pressure.

"Welcome," she said with narrowed eyes.

"Let's not tarry a moment longer, eh, chum? No rest for the wicked." He slapped Niko on the shoulder. Niko flinched, looking at him with distrust. Leonard grinned and headed off toward the end of the hall, leaving the rest of them bewildered.

"That's not Leonard," Niko said quietly.

"No shit, Sherlock," Lilian said.

"I feel I should point out he's got the only working stun gun," Allison added.

Niko nodded. "We'll have to snatch it off him."

Turning at the door, Leonard smiled warmly back at them. "What's the hold up, boys and ghouls?"

"Nothing, brother," Niko said with a nervous chuckle. "Just making plans."

Leonard clapped his hands together cheerfully, an oddly dainty gesture for such a tough-looking dude. "Wonderful! I love a good plan." He daubed the Kleenex to his nose and gave the blood a look of disgust. "Shall we split into pairs? Of course, we appear to be uneven so one of us will have to play the third wheel."

"I think we should stay together," Lilian said. "Right, Ben?"

Ben looked up as if he hadn't heard her. He nodded. There was blood on his lip. "Yeah," he said. "Whenever people split up in horror movies somebody always dies."

Leonard looked confused. He blinked hard and shrugged. "Very well, together we shall remain. I believe I know a faster route to the exit, anyhow." He turned and stepped briskly into the next corridor with the gait of an English butler.

"Soon as he lets his guard down we rush him," Niko whispered.

"I don't think that's a good idea," Allison said. "He could be dangerous."

Niko gave her a smug look. "I can deal with Leonard. We've scrapped a hundred times."

"But that's not—" Allison began, but Niko was already rounding the corner. She scowled and followed him.

Lilian heard the stun gun discharge a split second before Niko came lurching back into the doorway, his entire body seizing. He staggered back and toppled over a gurney, knocking it to the floor.

Leonard had grabbed Allison and spun her around. He held the stun gun to her head, the crook of his elbow pressed against her throat. Blinking angrily, he spat out a mouthful of her hair before attempting to speak. "No sudden moves, children, or the dame gets it."

"Let her go!" Lilian said.

"I'm not much for sharing, unfortunately. I'd much rather play with this one on my own."

As he spoke a drop of blood spilled from his nose and spattered on

Allison's forehead. She flinched and whimpered.

"Not another peep from you, my dear," the man who'd possessed Leonard said. "Or you'll get the shock treatment too. Ta-ta, children."

He raised the stun gun and twiddled his free fingers. Blinking repeatedly, he backed Allison down the dim corridor. Her shoes scuffed on the tiles until the two of them disappeared around the next corner.

DR. DEATH

BEN RAN FULL speed down the hall, chasing Allison and the ghost possessing Leonard. Lilian helped Niko to his feet and they followed him deeper into the asylum, their footfalls echoing down the long dark corridor.

Lilian caught up with him in the front foyer and they rushed up the stairs two steps at a time to the second floor. From there they followed trickles of Leonard's nosebleed on the black-and-white checkered tiles to a set of double doors leading into another long hallway. The overhead lights flickered and buzzed with crackles and arcs of wild electricity. Here, Ben took a moment to catch his breath. They had to wait on Niko anyway, who was climbing the last few steps holding the railing, puffing tiredly. Directly ahead of them was a lime green metal door with a small porthole-like window just above eye level. The trail of blood stopped below it. A large bloody handprint streaked the metal push plate.

The hallway to the left of the doors was blocked by a pile-up of roller coaster cars that had caught on a portion of track torn from the floor and bent sharply upward. The tour guide, dressed in a red Ghostland T-shirt and black Ghostland vest, was impaled on the jagged metal and badly burned, having closed the circuit on the electrified track before it must have shorted out. A few broken bodies lay beneath upturned tramcars, burning like guttering candles or smoldering like the tour guide. A fire extinguisher lay nearby, its hose loose, as if someone had already tried to put out the blaze. The smell made Lilian gag.

She stepped cautiously over the tracks and peered through the wired-glass window into what looked like a recreation room, the walls painted a soothing baby chick yellow. Melancholic piano music trickled out from within, one of the songs her music teacher used to play between students while Lilian waited in the foyer for her lesson to begin. The dead lay draped over sofas and in chairs, another under the table, more sprawled on the floor. Books and games had been pulled from the shelves, their pieces scattered. Lilian recognized the group of Japanese teenagers she'd seen taking a photo with Demont at the park entrance. They held hands, their red-rimmed eyes wide in horror, white foam oozing from their mouths. Lilian remembered what Allison had said about flash mobs and wondered if these kids had been a part of some suicide club.

She waved Ben over. "Come look. Watch the tracks."

Ben approached the window in the second door and squinted through. His eyes widened.

Niko stepped in behind them, panting. "Hey, guys. You're alive."

For now, she thought. She looked into the room at the scattered corpses. Something had killed those people and there was no reason to expect whatever it was wouldn't return. It could still be here, waiting for the right moment to pounce. Every part of her screamed at her to leave this place and never turn back.

But they couldn't keep running forever. They needed to find Allison. They needed that stun gun. They needed the code for the security hatch.

She pushed open the door, checked both ways before stepping through and then moved directly to a door beside a glass passthrough, where she guessed staff would hand out patient meds. The handle wouldn't turn so she walked up to the counter. On it was a clipboard with a checklist of names and medications: *lithium carbonate, chlorpromazine hydrochloride, imiprimine*. Beside the clipboard was a display with pills of various colors and the illnesses they would have treated.

The asylum was huge. They needed a clue to who they were dealing with, to whoever had possessed Leonard, if they hoped to find Allison alive.

Lilian pushed the pills aside and leaned over the counter. Within the small room was a medicine cabinet with glass doors, a wheelchair and a large rolltop desk. She jerked her head back out, watching the bodies on the floor, suddenly certain she'd seen one of them move in her peripherals.

"Anything?" Niko asked.

"It looks like a nurse's station," she said, returning her attention to the passthrough. "Just looking for anything we can use. A scalpel. Anesthetic. Something."

"Smart," Ben said. "I'll check the trash cans for money and ammunition."

Lilian laughed in spite of herself and climbed through to the nurse's station.

Niko chuckled. "Nerd humor. I get it."

Inside the medicine cabinet were some rusted tools that looked sharp but nothing that would be easy to use: a corkscrew-looking thing with a large wooden handle, a metal spatula, calipers, a rusty speculum. It was locked anyway, and breaking the glass would make too much noise, likely to alert whatever had killed the others. Discouraged, she crossed to the rolltop desk and tried the lid. It wouldn't open, no matter how hard she pulled and grunted.

"What's going on?" Ben asked.

She unlocked the door and stepped aside to let him in. "The lid's jammed," she said, nodding toward the desk.

He glanced at the cabinet. "Whoa, is that a skull drill?"

"Gross. Just help me with this, okay?"

Together they pulled on the brass handle.

"Need a hand?" Niko asked, leaning up against the doorjamb.

"We've got it," Ben said with undisguised hostility.

On the third try, the lock snapped and the lid rolled back into the desk. It was stuffed with stacks of yellowed paper and old ledgers. While Ben crossed to the medicine cabinet, Lilian picked up a leather-bound book at random and flipped through its musty pages, hoping to figure out who'd possessed Leonard, or come across a case history for Morton Welles, in case their paths crossed again.

No luck. The book contained the patient history of a woman named Emma Lou Amesbury[1], a former nun believed by the members of her convent to have been possessed by the devil. The words "violent sexual proclivities" had been repeated on several pages. Tucked within were several black and white photographs. In one, a sad, plain-looking woman in a long plain dress and frumpy bonnet sat in an uncomfortable-looking wooden chair. In the next, the same woman's head was wrapped in bandages between two glass plates on black metal poles. The third was a snow globe with a Christmas tree inside, the words *O Tannenbaum* etched on its base.

"I bet this is the patient in that room," she said. "If we could just find her snow globe—"

Ben slammed the book closed, raising a cloud of dust. She scowled at him.

"This isn't a *game*, Lilian. There aren't any trophies for completing side quests. There's nothing that'll help us in here. We need to find Allison before something happens to her."

Lilian looked at the books. The urge to take them with her was strong but there were far too many to carry. They wouldn't be of any use outside of the asylum anyhow, aside from the one she was least likely to find.

"Fine," she said, and she followed him out of the room.

Niko stood with his big arms crossed over his chest. "Where to now, guys?"

Ben pointed to a door at the far corner of the recreation room. "They must have gone that way."

Niko led the way, weaving between the bodies on the floor. "Such a shame," he said. "They're just kids. Who's gonna tell their parents?"

Lilian didn't know what to say. She watched her feet instead, trying not to look at the bodies. A cell phone lay in her path. She glanced at it just long enough to see it was playing a short handheld video of the dead Japanese kids hugging, singing, laughing, flashing peace signs and making duck faces. She turned away as the video looped back to the beginning and replayed.

NIKO DREW open the door at the far end of the room and peered inside. "Coast is clear," he said, holding it open. Lilian stepped into the cramped doctor's office. Ben and Niko crowded in behind her. The office belonged to a DR. HAMMERSMITH[2] according to the nameplate on the large oak desk and the framed diploma beside the bookshelf. Several drawers had been left open, and notes were scattered across the desk. Drops of blood had spattered the blotter and the floor leading to the far door.

Leonard and Allison had come through here.

Lilian paused a moment and looked over the photos on the walls, each one featuring a balding man in doctor's whites with round wire-frame glasses.

In one, he stood over a man receiving electroshock. In another, he stood beside a strange metal tank, a woman's head stuck out of the single hole, her mouth open in anguish. In a third, he held his arms behind his back among dozens of patients standing out front of the asylum. Placed in intervals between these photos were newspaper clippings, also framed. One headline read *Dr. Death Murders 18 Patients.* A second read, *Dr. Death Sentenced to Electric Chair!*

"Dr. Death," Ben said. "I wonder if that's who possessed Leonard?"

Niko balled his hands into fists. "Whoever it is, I'm gonna perform an exorcism with my fist up his ass."

A grainy photo of a familiar-looking man on one of the clippings stood out to Lilian. She stepped around Dr. Hammersmith's desk to get a closer look and her heart began to race. It was *him*. Younger, far less maniacal and lacking the lobotomy scar on his forehead, but it was definitely Morton Welles. Above his photograph, the headline read, *Dr. Death's Lobotomized 'Zombies' Claim 7 More Victims.*

"Morton Welles was one of his patients," she said.

Ben came around the desk to join her and scanned the article. "Whoa. That's cool."

She glowered at him.

"What? It is pretty cool, if you don't think about how twisted it is."

"This door's unlocked," Niko said behind them.

The other door led back into the hall. Lilian hurried past Niko, following the drops of blood to a set of double doors at the far end. The green copper plate beside the doors was labeled OPERATING THEATER, and a dark realization overcame her: Allison was behind those doors with Leonard, possessed by the sadistic Dr. Hammersmith, and the reason why Dr. Death had brought Allison here made her skin crawl.

"*He's gonna make her into a zombie,*" Ben said, echoing her thoughts.

"Not if I have anything to say about it," Niko grunted.

Lilian pushed hastily through the doors, no plan, just hoping she could convince the ghost occupying Leonard's body to let Allison go. She knew the chances were slim, but she couldn't just let Allison die. Though she'd never admit it, after all they'd been through today, she was actually starting to like her. The woman had forced her to face some hard truths, and if the park hadn't gone into meltdown, bringing her here might actually have helped. She would be sorry to lose Allison, no matter their differences in the past.

The room she stepped into was dim, three rows of seating arranged like a movie theater on a steep slope facing a bright room behind wire-frame glass. A door opened to the operating theater at the far end, with stairs leading down.

She took another cautious step inside until she could see the operating room below. Everything was bathed in bright white light. Leonard stood over a gurney, holding a scalpel. He wore a pair of horn-rimmed glasses, the same pair Dr. Hammersmith wore in the photographs. Allison lay on the gurney. She appeared to be immobile but awake, her eyes wide in terror.

A tray of surgical instruments lay between them. Leonard leaned over her, removing a white facemask from Allison's mouth and nose and stooping to turn off the gas tank at the foot of the gurney. He daubed sweat from his forehead with the bloody tissue, said something Lilian couldn't hear and cocked an ear toward his victim's face as if to listen for a response he clearly wouldn't get.

Allison's eyes bulged as his face neared hers. The doctor rose again and smiled, seemingly pleased.

Neither seemed to notice Lilian and the others had entered the theater. Lilian guessed it would be difficult to see up here with the difference in light, which gave the three of them the advantage of surprise. But it would be difficult to get down into the operating room without alerting Dr. Death.

"I'm gonna head for the door," Niko whispered. "When I get there, you two distract him."

That might work, Lilian thought. She and Ben nodded and Niko began creeping around low, keeping as much in the shadows as his massive frame would allow. Lilian stayed back with Ben, watching and waiting for their cue.

Dr. Death reached for Allison with the scalpel. Lilian saw her therapist's eyes follow the blade, which shimmered under the hot lights. Allison squeezed her eyes shut as he jabbed the blade into her shoulder.

"Did you feel that?" the doctor asked in Leonard's voice. Allison blinked hard but didn't reply as a rivulet of blood trickled down her arm. "No, I don't suspect you did. I've injected you with what's called 'twilight sleep,' a wonderful concoction of morphine and scopolamine. It's most often used for childbirth but I find it makes my patients quite... *pliable.*"

Leonard placed the scalpel on the tray and picked up what looked like an antique hand blender. He held it out for Allison to see. "As a student of medicine, I'm sure you're familiar with this instrument."

Tears spilled from Allison's eyes, shimmering in the light of the surgical lamp. Unable to move, unable to speak, she blinked hard once.

"What is that?" Lilian asked Ben.

"A bone saw," he said. "We have to stop him."

"Not yet." Lilian pointed toward Niko, whose shadowy form approached the door on the opposite end of the theater.

"He's gonna *cut her open*, Lilian."

"Come on," she said, heading cautiously down the steps toward the glass.

On the other side of the operating theater, Niko waved. That was it: the signal. Lilian began slapping her palms hard against the window. Blood streaked on the glass, the sudden sharp pain reminding her of her injuries from the midway. Ben stood beside her mimicking the action, shouting, "Hey, Dr. Dickhead! Hey, numbnuts!"

Leonard and Allison both looked up at the windows. The sorrow in Allison's eyes made Lilian's well up in sympathy.

Dr. Death grinned with Leonard's lips. He peeled off his glasses and wiped them on the breast of Leonard's uniform, blinking rapidly at the change in focus. Then he put them back on his nose and gave his guests a pleasant smile. "Ah, wonderful! It's been quite some time since I've had an audience."

On the other side of the operating room, Niko opened the door a crack. Ben pounded on the glass, whooping loudly. It was almost cathartic, Lilian thought. Whatever had happened to Ben downstairs, beating on the window would be a good destresser.

"Hush now, children!" Leonard said. "The good doctor has an operation to perform. I'm sure you wouldn't want me to accidentally sever an artery."

He flashed a toothy grin and flicked on the saw. Its two rusty, shovel-like blades whirred to life with a deafening whine. Leonard's eyes twinkled gleefully behind the glasses, and in that same moment, Lilian realized it wasn't just Dr. Death inside of Leonard.

Rex Garrote was in there with him.

Niko crept down the stairs. Lilian wanted to warn him. If he didn't reach them in time Garrote would kill them both, she was sure of it. Dr. Death was already smoothing back Allison's hair from her brow and lowering the whirring blades toward her forehead. Allison watched them, her eyes going cross, tears spilling down her temples onto the white pillowcase.

"Hey, Garrote!" Lilian shouted.

Leonard looked up at her, seething with rage. As he did another face flashed over his, as if she'd made the ghost angry enough to momentarily leave his host's body—but it was Garrote's face, not Dr. Hammersmith's. She was right.

Niko continued toward them. The sound of the blade masked his heavy footfalls, but he was still too far away. They had to keep Garrote distracted just long enough for Niko to take him down.

"That's right!" she shouted. "We know all about your virus!"

"Virus?" Leonard grinned, still affecting Dr. Death's mild accent. "Why, whatever do you mean?"

"Your code!" Ben said beside her. "We know you've been taking over the other ghosts."

Leonard seemed to consider it. Then he shook his head. "Poppycock!" he said and brought the blades down sharply on Allison's forehead.

"*No!*" Lilian cried, pounding feebly on the glass.

Niko jerked the stun gun from Leonard's belt as a fan of Allison's blood streaked his face. He jammed the weapon between Leonard's shoulder blades and zapped him. But the sawblades had already done their job. Allison had closed her eyes against the curtain of blood washing down her face while Leonard's body began to seize. The shock had made his body rigid, and his finger was frozen on the bone saw's trigger.

Bone, Lilian thought, looking at the white dust rising in a cloud around them. She turned away, unable to look, as the blades crunched through her therapist's skull and sliced into the meat of her brain.

MULTICOLORED LIGHT FLASHED through the glass partition. Lilian dared another look down just as Leonard wheel around with the bone saw still spinning, spattering blood everywhere. She saw Niko throw up his hands and she closed her eyes again, hearing the horrible sound of the bone saw tearing through raw meat and sinew, hearing the big man's shrill scream.

"Niko!" Ben ran past her to the other side of the theater, heading for the door.

She chased after him. "Wait! He'll kill you!"

Ben had already rounded the corner, shouting Niko and Allison's names. He tore open the door and dashed down the steps two at a time. Lilian hurried down behind him, not wanting to look at the grisly scene in the operating room but unable to stop herself.

Niko lay on the floor, drenched in blood, holding up his slashed wrists. Blood had poured over Allison's face and her scalp hung loose from her glistening skull, draped over the back of the gurney. The saw had gouged a black channel out of the bone above her eyebrows. If she survived the wound, she would likely be better off dead.

Leonard stood covered in Allison and Niko's blood with the bone saw still whirring in his hand. He startled, like a man waking from a dream, blinked away blood from behind the gore-streaked glasses and cast them aside. He turned to Niko, looked at the bone saw in his own hand and dropped it in horror. It clattered on the floor beside Dr. Hammersmith's glasses, bits of brain and blood-soaked skull spattering away from it.

"What the fuck?" he said. "*What the fuck, Niko?* Did I fucking do that to you?"

Ben kicked the plug out of the wall and the saw slowed to a stop. He hurried to Niko's side. "We need some bandages," he said, unzipping his backpack and rooting inside.

"Leonard, brother. It's not your fault. Don't freak out."

"I..." Leonard turned. He saw Allison dead on the table and shook his head madly. "Oh no. No, no, no! All I remember is being in that room, those footprints... there was a man... this doctor... Oh, Jesus, I can't believe this. I can't fucking *believe* this!"

"Just get some fucking towels!" Ben snapped. He'd unzipped his bag and was pressing folded tissues to Niko's slash wounds but blood had already soaked through them.

Leonard nodded sheepishly and got into action. He tore off his sleeves, exposing tanned, freckled arms and pale shoulders, then brought the rags to Niko and knelt beside him. He handed one to Ben and the two of them began tying them around the deep, oozing gashes on Niko's forearms.

Lilian watched the scene in a daze. It didn't seem real. Time had slowed down. It felt like she was dreaming, or stoned. She approached the woman on the gurney with the agonizing sluggishness of a dream, while behind her the men kept shouting at each other, their voices muffled by the strange soupiness in Lilian's brain. Allison's glassy, lifeless eyes stared up at the ceiling. Lilian tripped over the bone saw, and it clattered against the foot of the gurney.

Allison flinched at the sound.

"She's alive!" Lilian cried.

She vaguely heard someone say, "Thank God." It could have been any of them.

The therapist's gaze rolled to the side, toward Lilian, but not finding her. "Lilian?" she said. Her voice was soft and childlike, like someone muttering in their sleep. It was nothing like the woman she'd known, as if the injury had stripped away the years of polish and education and psychological armor she'd used to navigate the world and left her with just the shell of herself, a blank-slate Allison Wexler. The glistening, blood-red flap of her scalp flopped as Allison frowned, swallowed and spoke again. "Lilian?" she said, reaching out blindly.

Lilian took her hand and squeezed it. "I'm right here," she said.

The therapist whimpered, her mask of blood creasing at the corners of her lips. "I can see him down there," she whispered. "He's waiting for us in the dark." The words prickled the hairs at the back of Lilian's neck. "He's in the water. Oh God, it's *so red.*" She squeezed Lilian's hand hard. "Why is the water so *red*?" Allison's eyes, flecked with small drops of her own blood, locked on Lilian's. "Don't go down there, Lilian. Promise me, you won't go down there in the dar—"

Allison's whole body shuddered suddenly, and her voice trailed away on a final breath, her chest deflating under the hospital-blue sheet.

"Dr. Wexler?" Lilian said, and before she could stop herself, she sobbed, stifling it behind a hand.

Allison was dead. Allison was dead and Lilian had no one to blame but herself.

The same would happen to each of them, one by one, until there was no one left to weep over their unhallowed graves. Garrote would drag them all down into the dark, into the red, red water. Lilian had no idea what the words meant but an image flashed so clear of Allison and Niko and Leonard and Ben, all of them splashing madly in a river of blood in the dark, gasping for air as the disembodied hands of Garrote's ghosts dragged them under.

She turned to Ben and the others, trying to shake the image. They were still wrapping up Niko's wounds. God, there was so much blood! Turning back to Allison, she tried to put the woman's last words out of her mind. Her spirit would be crossing over soon, and Lilian wanted to help her the way Allison had helped the man at the farmhouse. Allison had been so calm, so helpful—and for the first time, Lilian had truly seen her therapist shine.

But they had to leave. They couldn't stay here much longer. Garrote's power grew stronger by the minute. Sooner or later, he would stop playing games and with every ghost in the park under his control the survivors—now only four of them—wouldn't stand a chance.

She let go of Allison's cold hands. Without taking her eyes off her therapist's face, frozen in a nervous smile, her fingers found the bracelet her parents had given her. She unsnapped the clasp and slipped it off, then clasped it around the cool skin of Allison's wrist. The jewelry didn't mean much to her but she hoped it might act as a totem for Allison when her spirit left her body. That she'd understand what it meant and know what she had tried to do for Lilian meant something.

Lilian brushed the dead woman's eyes closed. "We're gonna stop him, Allison," she said, not sure she believed it, only hoping to be a comfort if Allison could somehow still hear her. "Your death won't be in vain."

She smiled, briskly wiped away a tear, and turned back to the others. They were still kneeling beside Niko. His face had gone ashy gray. The torn fabric they'd tied around his injuries had already soaked through with blood.

Lilian stooped and picked up the stun gun. She twisted it, saw the power light was still green. They had a weapon again, something to protect them against Garrote and his ghosts. But she couldn't trust Leonard, not yet. She had to be sure. Creeping up behind him as he kept pressure on the torn fabric around his friend's wound, she prepared herself mentally to pull the trigger. She knew how much these things could hurt. She'd seen videos on YouTube.

Ben saw her approaching and his eyes widened. He must have seen what she was planning because he mouthed "no," shaking his head.

Leonard looked up, his eyebrows rising in comical surprise.

"Don't move," she said, her voice unsteady. The stun gun quivered in her grip.

The big man held up his hands. The palms were covered in Niko and

Allison's blood. It wasn't his fault, she had to remember that. But whoever had done this to her friend could still be inside of him. It wasn't worth risking another attack.

"Lilian, he's gone," Ben said. "Dr. Death—Niko already zapped him out of him."

"But what about Garrote?" Lilian said. "How can we be sure he's gone too?"

Ben's mouth opened in surprise. He turned to Leonard, suddenly suspicious.

"Hey. Hey, little lady." Leonard looked up from Niko's side with sad puppy dog eyes. "If it'll make you feel any better I'll let you stun me. Just not while I'm holding onto my buddy's arm, okay? Chain reaction and all that."

Lilian nodded for him to let go. Leonard stood, raising his hands in surrender. "Okay now, do you know how to use—?"

She jabbed the stun gun into his chest and pressed the trigger. Leonard's face contorted and his limbs began to jitter. He fell back flat on the floor and his legs twitched like a dying bug's, like he was breakdancing.

"*Jesus*, Lil!"

When his limbs finally stopped quivering and fell back to the floor with meaty slaps, he sucked in a deep breath and exhaled sharply. "God*damn* that smarts!" After a moment he pushed himself up with difficulty and gave her a hard look, rubbing where she'd stunned him. "Are we cool now, chica?"

"For now." She spotted a door in the far corner of the operating room that led beneath the theater seating. "We should get moving," she said, scanning the dark theater above them.

Anything could be lurking up there in the shadows—Morton Welles, or another of Dr. Death's zombie patients. The evil sex nun, Emma Lou Whatshername.

"Think you're all right to walk, buddy?"

Niko nodded. "Just help me up," he grunted.

Ben and Leonard slipped their hands under Niko's arms. He groaned as they helped him to his feet. "You cut me deep, Shrek," Niko said to Leonard, who chuckled bitterly.

"Good to see you kept your sense of humor," Leonard said.

Walking on his own, Niko staggered toward the door. Ben, Leonard and Lilian followed.

"She was a good woman," Niko said, stopping at the end of the gurney to look down at Allison.

"I am *so* sorry," Leonard said, favoring Lilian with a sad smile. "She didn't deserve to go like that."

Lilian said, "It's not your fault."

"I could've fought him off. It was like I was watching myself in a dream. I coulda stopped him!"

"Nobody's blaming you," Ben said. "I just wish we could help her." He eyed the bracelet Lilian had put on Allison's wrist, sniffled and gave Lilian a morose look, wiping his eyes with the back of a hand. "You know, with her transition. It doesn't seem fair just to leave her like this."

"I think she'll do all right," Lilian said. "She's a tough lady. It's Garrote getting inside of her I'm worried about."

Ben nodded. "If that happens, we have to stun her. We have to promise we'll do that for each other, too. Before he gets into us."

Each of them made a promise. Leonard walked ahead and held the door open for them. They all piled out into another dim corridor.

"Now what?" Niko asked.

"*Hello?*" A male voice came from their right, followed by a blast of static. "*If anybody can hear me, please respond.*"

Static followed.

"That voice," Ben said, scowling down the hall. "It sounds familiar."

Lilian recognized it too. "Demont!" she said. He'd come back for them. Maybe there was a chance after all. Without another word, without thinking, she ran in the direction of his voice.

Away from death, Lilian ran toward hope.

VOICES FROM BEYOND

BEN WATCHED LILIAN run off toward Demont's voice and for a moment he was too stunned to follow her. He'd never seen someone he cared about die before, had never seen a dead person before until today, and the shock of watching Allison die was still raw, like the wounds on Niko's forearms. And now, without a word of warning, Lilian had run off alone. She was going to get herself killed.

And it's all my fault, he thought.

He turned to the others. Leonard was propping Niko up and fiddling with his walkie-talkie. Blood had already spattered the floor around Niko's feet in the few moments they'd been standing there. His face was gray and his pupils were very small. Leonard caught him as he swayed. He needed medical attention as soon as possible. Ben remembered Guest Services had been equipped with several medical cabinets but they were still a good distance away from there. He doubted they would make it in time.

"Come on, guys," he said. "We need to catch up."

Leonard ignored him, twisting the dial on his walkie. "He must be using a different frequency."

"Maybe he's on the asylum intercom," Niko groaned.

Leonard shook his head, concentrating on the dial. "I know the sound of a squawkbox when I hear it. I haven't heard a damn thing on this since the shit went down, but it could just be anyone who survived is caught somewhere and needs to stay quiet. Or they had bigger fish to fry, injuries and such."

Ben didn't care, frankly. Demont was out there, which meant there might also be other survivors. Joe the mechanic could be there with him, or someone else who might have access to the security hatch code. "I'm gonna go find Lilian," he said.

Leonard nodded, still fiddling with the dial as if it made any difference. The thing was probably broken. "We'll catch you up," he said, distracted.

Ben stepped out into the hall. The next corridor appeared to be another patient wing, with the same lime-green doors recessed into the walls at regular intervals. It sloped downward to the left. A sign with an arrow pointed that way, labeled MORGUE. At the opposite end, Lilian stood looking into an open doorway. He'd expected her to have gotten a lot farther by now. He called out her name.

She turned, fear showing on her pale face even from a distance.

With his back toward the morgue, Ben glanced into the room to his immediate right. The walls had cracked and the mattress had been flipped off the bed. A woman—at least he thought it was a woman—sat in a chair in the corner of the room. Her jaw had been carved off and lay in her lap on a blood-stained white sheet. She held a scalpel in her right hand and she jittered uncontrollably, blood and saliva dripping from the wound and from the scalpel and the hand she held it with, her uvula lolling as she made glottal sounds as if she were attempting to speak.

Ben looked away and hurried past the room. As Niko and Leonard finally shuffled into the hall, Leonard looked into the room and gagged. "Jesus, what the fuck!"

Through the next open door, frantic spirals were being drawn in charcoal on the walls, the bedspread, the floor. The overpowering cooked onion stench of bad body odor wafted out of the room but other than the levitating charcoal there seemed to be no physical entity within.

Ben held his nose and moved on. Behind the next a wiry elderly man with a straggly gray beard and a diaper around his waist stood beneath the caged window, digging fingers into a raw hole in his stomach and pulling out loops of intestines that splatted wetly on the floor at his feet. The room smelled of feces and sour cabbage.

As he rushed past several more doors, consciously avoiding looking inside, another sharp burst of static made him jump, and Demont spoke again. Leonard had been right. No mistaking that tinny walkie-talkie sound. "*This is Demont Hudson. I'm a Suicide Prevention Officer. If anyone out there can hear me, please respond.*"

Wherever the walkie was, Lilian made no move to answer the call. She was still staring into the final room at the end of the hall.

"Lil," he said sharply. "What's wrong?"

Lilian raised her hand very slowly and put a finger to her lips. He came to her side, close enough to feel the warmth radiating from her. He hadn't noticed how cold it had gotten in here until just then. When he looked into the room, he saw why. Fresh snow had powdered the floor. A man in boxer shorts lay propped up against the wall. The nun knelt between his legs, the heart shape of her bruised, bare ass, slashed with angry red strap marks, facing the door as she leaned over the man's groin, working with her hands between his legs.

Between this macabre scene and the door lay the walkie-talkie.

"*Ah, forget it*," Demont muttered to himself. "*There's nobody out there.*"

Static. Then nothing, just the sound of rending flesh as the nun severed Joe the mechanic's manhood.

Ben wanted to make a mad dash for the walkie but fear had frozen him in place. He could still feel the phantom pressure of her breasts against his chest. He could still taste her rancid breath and the sour apple mush she'd forced into his mouth with her tongue.

Her shy boy.

"Oh, shit—Joe!" Leonard cried, standing next to Ben.

"Is he...?" Niko lurched up to the doorway and dropped back against the wall opposite with a groan of exhaustion. "I gotta sit down," he said. Ben heard the man sit on the floor behind him but he did not turn around. He *couldn't* turn around. His eyes were fixed on the woman in the room.

The nun pulled back her right hand with a squelching sound and came up with a bloody fist full of wriggling genitals. She turned her head to the side, holding up the man's severed penis like an offering to God, and let the blood drip off its tip onto her curled tongue.

Leonard gagged. "Ugh, I think I'm gonna *puke*."

The nun bit the head off the penis and chewed, her lips smacking wetly.

True to his word, Leonard hurried off to the side, planted his hands against the wall and splashed vomit between his boots.

The nun's black eyes twitched toward the doorway.

Ben flinched but didn't move, held by her dark gaze.

"*Oh my, my, my...*" The nun licked blood from her lips. "*My. Shy. Boy.*"

Lilian turned to him. He felt her studying him but he didn't react. He couldn't look away and he couldn't speak. All he could do was watch, despite not wanting to see, while the nun chewed on the mechanic's grisly member until she swallowed hard and uttered a satisfied, "*Ahh!*"

"We need that walkie," Lilian said, still eyeing him.

"I can't..." Ben shook his head. "I can't go in there."

"Did she do something to you?"

He opened his mouth to reply and swallowed the nun's putrid taste. He couldn't tell her. What would he say? It didn't matter anyway. All he could manage was a nod.

Lilian stepped forward, one foot into the room. Ben snatched out and grabbed her arm.

"*Don't.*"

"I'm not her type," she said, shaking off his hand to raise the stun gun. "This bitch is mine."

The nun turned and stood in one lithe movement, planting her bare feet apart, standing now between Lilian and the walkie-talkie. Drenched to the elbows with blood like she was wearing fancy red velvet gloves, her fingers dripped as she wiped her lips with the back of a hand and showed her rotten, blood-pinked teeth in a smile.

With her attention diverted, Ben glanced at the man between her legs. The boxers had been jerked down below his ass. His hairy belly sagged over his lap but enough of his groin was visible to see the black, gaping hole oozing yellow fatty tissue where his genitals should have been. Combined with the stink of Leonard's puke, Ben nearly threw up himself.

"How'd you like a *mazel tov* cocktail, bitch?" Lilian said. It was one of her favorite taunt phrases from when she and Ben used to game together. But even nostalgia couldn't wash the stain of sin from his lips.

The nun's head twitched within the cowl, her black eyes watching Lilian's cautious approach. "*The shy one belongs to you, is that so?*"

"He doesn't belong to anybody," Lilian said, fear quavering in her voice. "He's my friend!"

"*Pity you didn't have him yourself. Virgins make such delightful little playthings.*" The nun grinned, her teeth slick with the mechanic's dark blood. "*What's the matter, shy-boy?*" Her voice had deepened to a gurgling, almost masculine growl. "*Pussycat got your tongue?*"

The ghost roared and charged at Lilian, her jaws opening impossibly wide, a gaping black hole filled with razor-sharp teeth. Lilian pulled the trigger as the nun pounced, rising off the floor. The charge struck the dead woman between her jostling breasts and her dark eyes bulged in surprised terror. She howled. Her form glowed a brilliant white for a moment, her head and limbs twisting and jerking, and in that frenetic moment, her face became Garrote's face and he howled in impotent rage, his ghostly limbs stretching out from the nun's shuddering body to grab at the arms of his destroyer.

Lilian stepped back out of his reach. In the next moment Garrote and the ghost he'd infected burst in a flurry of snow that cascaded down over Lilian's head and shoulders and the floor at her feet.

After a moment Lilian turned, breathing heavily, a ghost of a smile on her lips. The snowflakes had already melted, no trace of the nun or Garrote left in the room. "That was kind of awesome," she said with an anxious laugh. "I don't think you'll have to worry about her anymore."

"I'm sorry," he began, wanting to explain his fear to her. "I couldn't—" He shook his head again, unable to put it into words that wouldn't mortify him. What had happened was far worse now, knowing Garrote may have been controlling the nun from the very beginning, that he hadn't just had her rotten tongue in his mouth but the writer's as well. That he'd had Garrote's filthy tongue down his throat.

Lilian hugged him. He hugged her back weakly, wary of the sticky stiffness below his hips.

"Heckuva job, little lady," Leonard said behind them.

Lilian let Ben go and bowed theatrically. She tucked the stun gun into her back pocket and stooped to pick up the walkie. Raising it to her lips she pressed the button, then let it go again, hesitating. "Wish me luck?" she said.

"Just call him already," Ben said, huffing in annoyance. The quicker they figured out their next move the sooner they could get out of this fucking awful place.

Lilian pressed the call button and held the walkie to her lips. "Demont, this is Lilian Roth. Do you copy?"

A squall of static. Everyone waited, anxious, hoping he'd respond.

"Demont Hudson, do you copy?" she said, a touch of panic in her voice.

Nothing but static. She lowered the walkie with a groan of disappointment. Ben felt it too.

"*Demont here,*" the man said frantically on the other end. "*Boy, am I glad to hear your voice!*"

BEN COULD SEE Lilian was trying to keep her excitement in check but she was smiling from ear to ear. "Not as glad as we are to hear you," she said over the walkie. "Where are you?"

"*I'm at the prison,*" Demont said. "*Where are you all at?*"

"We're at the sanitarium," Ben said.

"*That's not too far from me. Think you can get here in one piece?*"

"We'll definitely try," Lilian said with a hopeful smile.

"*Great. I've got the place fortified. I'm pretty sure we could ride this out until help comes but... I need you to do me a little favor first.*"

"What?" Ben asked.

Lilian frowned at him. She pressed the call button and repeated the query.

"*There's an automotive museum, just a little out of your way. This generator's gonna conk out soon if I don't feed it some juice, and it's the only thing keeping this place safe. Think you could get some gas for me, Lilian?*"

She looked to the others questioningly. Ben nodded. He didn't think they'd be likely to find any gas even at the automotive museum but it wouldn't hurt to say they would try. Whether they did or not was still up for debate.

"Okay, we can do that," she said. "We'll be there soon."

"*Good luck,*" Demont said. "*Over and out.*"

Lilian jumped up and down, the buttons on her jean jacket tinkling like little bells of joy. "We're saved, we're saved, we're saved!" she cheered.

Ben wished he could be so certain. They'd be walking right through the front doors of the most dangerous, most haunted prison in the world. It didn't seem like a very smart plan at all to him. But Demont was there and somehow, he was still alive. That had to count for something.

"Hey, let's not throw a party just yet, little lady. We still gotta get there. Right, Niko?"

Niko didn't answer. The man was sitting against the wall in a pool of blood, his face entirely gray, his brown eyes staring at the ceiling, lifeless and unblinking. He was dead. First Allison and now Niko.

Who else, before this day is over? Ben wondered. *How soon until it happens to the rest of us? Will any of us make it out of this fucking place alive?*

"Aw, buddy..." Leonard crouched beside his friend. His knees popped and he grimaced, his eyes welling with tears. "Man, I am so sorry. It shoulda been me, bud." He closed Niko's eyes, shook his head. "All the times you saved my ass, it shoulda been me."

A moment of silence passed before Leonard looked up from his friend's lifeless body. "Let's go," he said. "Niko would've wanted us to carry on without him." With that he headed off down the hall in the opposite direction, hanging his head. Lilian gave a last lingering look at Niko and followed.

Ben hung back, feeling like he needed to say something. Regardless of all that had happened today, the man had given him a break letting him through the gate despite his medical condition. Like Leonard, he'd also saved their asses more than once. But Ben had betrayed Niko to get here. He'd lied to him by omission. The lighter fluid he'd sneaked in sloshed as he knelt beside the man.

"I'm sorry, Niko," he said. "I'm gonna make sure those two get out of here alive. Whatever it takes. My heart is strong enough, I promise."

He stood there a moment, not sure exactly what he was waiting for. Maybe he hoped Niko's ghost would respond to him. Make the lights flash. Jingle the keys on his belt. Anything. When nothing happened for another long moment, he unsnapped the keys from Niko's belt. It felt a bit like robbing the dead but he knew if they came across an instance where keys were needed and he hadn't taken them he would regret it. Then he stood, leaving Niko in silence.

"Where to now?" he asked as he caught up with Lilian and Leonard at the end of the hall.

"The exit's not far from here." Leonard indicated ahead with a nod. His eyes were red from crying. "Just down this next hall, I think."

"Great. Let's get out of here."

"Agreed," Lilian said.

They turned the corner. Ahead of them the tram tracks went under a set of double doors recessed in the green marble wall. A weathered copper sign to the left of the doors said HYDROTHERAPY.

Stopping there, Leonard said, "What I don't get is, how come I couldn't call your buddy up on my walkie?"

"I thought you said he was using a different frequency," Ben said.

"I thought that. But I searched through the whole dial while ya'll were talking." The ex-Marine shook his head. "I couldn't get a damn thing."

"Maybe there's something wrong with your radio," Lilian suggested.

Leonard looked down at the thing hanging from a clip on his belt and shrugged, then nodded. "Maybe." He pushed open one of the doors and held it for them. Ben stepped through first. Lilian entered behind him.

The hydrotherapy room was tall and wide, the ceiling a series of arches, the whole space tiled with green marble squares. It smelled like chlorine, reminding Ben of the school pool and the bullying he'd had to deal with in the locker room before he'd been pulled from school, where their gym teacher had never kept watch. In a way he was glad he'd never had to go back there, despite homeschooling having made him even more of an outcast. More than the pain, the beatings at the hands of kids much bigger and stronger than him had been a constant source of embarrassment. In a way, his heart attack could have been seen as a blessing, although he'd never think of it like that, especially after today.

The tram tracks ran between two large swimming pools filled with clear, aqua-tinted water, where the four connected tram cars had stopped. Like the ones they'd seen upstairs, each was large enough to seat four passengers. The space between the train and the pool was narrow, and several people were still seated in them, slumped over the bar and the sides, clearly dead, their clothes and hair damp as if they'd drowned sitting up in their seats. Very likely, they had.

"Awesome," Lilian said sarcastically.

"We could go around," Leonard suggested.

Water trickled and slopped in the pool and in the bathtubs along the walls. Steam rose from them like the remnants of dispersed ghosts, too thick to see what might lie beneath the blue-green water.

"I'll take my chances," Ben said. "At least we know what's in the train."

"I guess," Lilian said. She began edging her way along the right-hand pool like she was walking a tightrope, careful to avoid the dead within the train.

Leonard stopped at the edge of the pool and turned. Ben had been watching the water and almost bumped into him. "That girl is tough as nails," the man said. "Are you two, uh...? You know."

Ben sputtered to cover a blush. "Dude, she's practically my sister."

"Yeah, but she's not, right? Hey, man, I'm just askin'. If I was in your shoes..." Leonard let the thought linger, the implication more than enough.

Already standing on the other side of the pool, Lilian shot back a look of suspicion. "What are you guys talking about?"

"Nothing," Ben said hurriedly. He frowned at Leonard before the man could open his big mouth and sell him out.

"Well, come on then. I can't wait all day."

Leonard began across the narrow gap, walking surprisingly lightly for a man of his size. Ben followed a few paces behind, watching the bodies for movement. A man in a blue hat and a WWE T-shirt riding over his pale white gut was sitting upright in the back of the second car, trapped by the safety bar. His glassy eyes stared straight ahead, mouth open, tongue touching his lower teeth. His right arm hung over the side, close enough to reach out and grab Ben's legs.

Ben glanced into the pool as he sidled past. A woman in a pink dress floated face-down on a pool noodle near the ladder. Her dress had bloomed around her waist, revealing the black bicycle shorts she wore beneath. As he got closer, he realized it wasn't a pool noodle, it was a hairy leg, severed at the hip. The rest of the man's body was nowhere in sight.

"It's about time," Lilian said as Ben reached the other end of the pool.

Directly ahead of them an archway led into another high-ceilinged room. On the other side were several steam cabinets. Ben recognized them from research into old psychiatric treatments and torture devices. Each cabinet had a single hole for the manic patient's head to stick out of while their physicians steamed them like clams. It was meant to be calming. It wasn't.

A long, groaning creak came from one of the steam cabinets. Its doors opened and a man in a hospital-green gown rose from inside. He wore fuzzy slippers and held a cleaver. Ben recognized him immediately. Lilian grabbed his hand and squeezed it with a look of sheer panic.

It was Morton Welles.

Another cabinet swung open, then a third, releasing two more of Dr. Death's zombies, a man and a woman with the same prefrontal lobotomy scars on their foreheads, their lower lips drooping, their eyes rolled back as if searching for the portion of brain the doctor had cut out of their heads. The three zombies stood on either side of the door, making escape through there unlikely if not impossible.

Ben turned to Lilian. She was taking little panting breaths and reached for her bracelet with her free hand but the bracelet was no longer there, no totem left to calm her. He cupped the hand he held and squeezed it tightly, giving her a smile of encouragement even though he was just as terrified as she was, just as devoid of any hope.

"*Hewwo, Wiwian,*" the lobotomized zombie said in a mush-mouthed monotone, his tongue too fat for his mouth. "*Wememba me? I was cwazy too, but now I'm awwwww betta. You can be betta too, just wike me. Just wike Awwison. Aww you need is a wittle cut wight here,*" he said, dragging the edge of his blade across his own sutured scar.

"No," Lilian muttered, shaking her head fervently. "I'm not crazy. I'm not like you."

"Don't listen to him," Ben said.

Morton Welles turned slowly toward him, raising the viscera-stained knife. A little runner of drool spilled from his bottom lip as he spoke. "*Don't fink I fuhgot about you, Ben. We're gonna save you for wast. Oh yes, we arrre. We're gonna kill aww your fwiends and you're gonna get a fwunt woah seat.*"

As the zombie spoke Ben realized this wasn't Morton Welles, not anymore. The Garrote virus had infected it. The writer was speaking through him, threatening them, using the zombie to torment Lilian because Welles was her bogeyman, in the same way he'd used the nun, because she'd been who Ben had most feared.

Lilian had saved him from the nun. He would have to save her from Morton Welles.

Splashes from behind made her jump, and Ben along with her. She gasped. He didn't need to turn to know more lobotomized zombies were rising from the baths. This was their sanctuary, a place of the dead. The living were invaders. All invaders would die.

"We gotta get out of here, pronto," Leonard said.

Ben let go of Lilian's hand and began rifling through his backpack. He found the VR controller he'd picked up at the doors, switched it on so the plastic bulb glowed pink, and he lobbed it at Welles, like a miniature electronic hand grenade.

The controller struck the ghost right between the ribs, and the fabric of his grimy hospital gown and flesh beneath peeled back like burning paper. Welles looked down at himself, drool spilling from his lower lip and vanishing in the midst of the swirling vortex in his torso as the small electronic struck the tiled wall and the plastic shattered.

The two other zombies from the steam cabinets stepped back in fear. Ben grabbed Lilian and pulled her along into the next room, hoping Leonard would follow.

A large metal drain occupied the center of the sloped tile floor. Tall, curved pipes with shower heads rose from the tiles at intervals along both walls. At the far end of the room, something like a lifeguard tower stood where a mean-looking orderly, his hairy flesh blue and bloated, sat gripping the lever and large copper nozzle of a giant hose in his hairy-knuckled hands. Ben knew the pressure from the hose would push them back. Like the steam cabinets, it was meant to subdue manic patients.

Slippers slapped and wet feet squelched on tile behind them from the other room. Eight lobotomized zombies shuffled and staggered toward the shower room, drooling and dazed, flaps of rotten gray flesh sloughing off their bloated corpses like boiled meat.

They were trapped, closed in on all sides.

With a single stun gun between them, they were also outnumbered. No chance of fighting them off. They'd have to risk the frigid blast of the orderly's hose if they wanted to escape.

"Make a beeline for that door," Leonard whispered. "I'll try to draw this prick's attention."

"But... they'll kill you," Lilian said.

"I'm a Marine," Leonard said. "If I'm gonna die it'll be with my goddamn boots on—now go on, before I regret it!"

They nodded and hurried along the wall beneath the showerheads. Leonard stepped out into the middle of the room and began waving his arms. The orderly tracked him with the hose. "Hey, man!" Leonard shouted. "Hey, what'd they do, shave an asshole and stick it in a uniform?"

Sneering, the orderly pulled back the lever. The torrent struck Leonard and launched him off his feet. He staggered, leaning into the blast like a man facing a hurricane. Then he slipped in a puddle and slid back into the corner of the room, slamming into the wall.

Dr. Death's zombies shuffled into the room but the orderly didn't let up on Leonard. Every time the Marine tried to get up, the water pounded him back against the wall. Gargling and crying out, he raised his arms to protect his face and spat a mouthful of water. "That's all you got? I piss harder than that!"

Another blast made him scream. He wouldn't last much longer, and in the meantime the zombies had narrowed the gap, the first of them, Morton Welles, shambling toward Leonard's prostrate body with a large gaping hole in his torso.

Ben reached the door, Lilian just steps behind him. He grabbed the handle and tore it open. It wouldn't open far, the bottom wedged against the floor where it had apparently caught many times before, judging by the chipped and scraped tiles. Without a word, Lilian scrambled through the narrow opening.

Halfway out the door, Ben looked back at their savior. The zombies had swarmed him. The Marine caught Ben's eye briefly and waved for him to keep moving as Dr. Death's lobotomized patients slashed and hacked at his limbs, his chest, his throat, while the orderly's hose sprayed their blood and rotted flesh away in a pink-gray mist.

Ben pulled the door closed behind them, silencing Leonard's scream. He felt terrible, as low as he had ever felt in his life. He'd promised Niko he would get his partner out of the asylum alive and Leonard hadn't even lasted ten minutes. Allison and Niko and Leonard were gone and he and Lilian hadn't even gotten the door code to make their deaths worth something.

Every minute they'd spent in that hellhole hospital had been for nothing. *Less* than nothing.

He caught up with Lilian, who stood in the courtyard looking off at the surrounding exhibits. When she noticed him at her side, she grabbed him by the shoulders and pulled him into an embrace. They stood in the cold shadow of Bright Falls Sanitarium, weeping in relief to finally leave it behind, and in grief for the friends they had lost.

PART 3
PRISONERS OF THE DEAD

"Things fall apart, Father Brady," the Vampyre said. "It is the one thing you can count on. The only perfect thing in your god's world is Annihilation."

— Rex Garrote, *Ghost World*
Ep. 1.4 "Sins of the Blood"

Even when they aren't performing, ghosts are forced to endure endless torment from this abhorrent technology masquerading as "entertainment." Employees colloquially called "keepers" wear protective suits which emit static charges to "keep ghosts in line," according to an inside source. A more accurate description of this practice is "torture."

— #GRP2 pamphlet

ALONE

LILIAN AND BEN shuffled along the promenade, past silent exhibits and grisly tableaus of death. They hadn't seen or heard any sign of survivors since leaving the asylum, but they'd had to run and hide several times as a spectral creature shot out of a window or a darkened corner and shuttled away from them, heading northeast. They'd crept past a blood-soaked clown cheerily juggling the severed heads of his victims and a woman in a white flowing dress spinning the dead into cocoons made of living vines. And everywhere, corpses littered the ground, hung from trees, through broken windows, in burning trash bins, nailed to exhibit signs.

It was impossible to walk twenty feet without running into the recently deceased, and the bodies were beginning to ripen and draw flies. A sort of vague blue cheese-rotten vegetable smell permeated everything, like the smell that wafted up from the kitchen garbage disposal when Lilian's dad forgot to clean it for too long. She could even smell it in her hair, like cigarette smoke after a house party. It wasn't bad enough yet to make her feel ill but she knew it was the stink of the rotting dead, and she knew it would only get stronger as the day waned and night approached.

She thought they might have to consider making facemasks for themselves soon. If they stayed alive long enough to need them.

Worse, they had no idea where the Museum of Haunted Vehicles was located. For all the dead they'd seen, not a single map had been dropped nearby. She'd wanted to flick on the walkie-talkie and ask Demont but Ben was adamant it would be a bad idea. In horror movies and video games, the person on the receiving end was often in a tight situation when the call came in and it could get him killed. Lilian didn't think his logic was sound but she hadn't been in the mood to argue. He knew the vague direction it was in from looking at the map back when he'd still had it—he couldn't pinpoint when he'd lost the thing but it had likely been somewhere prior to the asylum, he'd said—and so they'd walked that way, weaving between exhibits, dodging stray ghosts and mass graves, until finally Lilian spotted a dead man sprawled over a trash can and pointed him out to Ben.

"What are we playing, I Spy?" Ben said. "I spy with my little eye, something that's starting to stink."

She swatted his shoulder. "No, you idiot. Look in his pocket!"

Ben looked closer and spotted what she'd already seen: a park map poking out of the dead man's back pocket. His crooked grin widened. "Holy shit! Nice work!" he said and started off across the road toward it. He stopped abruptly a few feet away from the man with a look of sudden realization.

"Well, go on and get it," Lilian said.

Ben reached out gingerly and tweezed the map between his index and middle finger, pulled it delicately out while eyeballing the man's dead-eyed, twisted grimace for movement. Then he hurried back to her side.

He flapped the map open and they studied it together. She found the cartoon icon for the asylum easily and traced her finger along the route they had followed. Their destination was close to Guest Services and unless the scale was off, getting there wouldn't take them too much longer but it had taken them quite far out of their way.

Their route planned, Ben folded the map and went to tuck it into his back pocket. He thought better of it, slipped it into a cargo pocket in his shorts and snapped the buttons shut. Then they struck off toward the Museum of Haunted Vehicles. For all the trouble it had been to get them this far, Lilian hoped to hell they would find what they were looking for, and that Demont's sanctuary inside the prison was as fortified as he'd said.

"I guess we're the only ones still left alive," Ben said after a long period of silence. "The last living ghosts."

Lilian gave him a sidelong glance. It was just the sort of weird phrase Ben used to say out of nowhere when they were still close friends. Funny to think he hadn't changed a bit in four years when she was so different. "That's random," she said.

Ben shrugged. "You know me. Rando Calrissian." He gave a weary smile. "It's from a Rex Garrote book. *Shōki*, it's called."

"I think we've all heard enough from that crazy fucking asshole today. I don't need you quoting him."

Ben apologized. They kept walking.

"I know it's weird, but I actually miss Allison," Lilian said after a time. "She was annoying as hell. But she was only ever trying to help."

"That's not weird," Ben said. "I miss her too. She didn't deserve to die like that. And nobody's going to pay for what happened, that's the worst part. They'll shut this place down, and maybe the company that runs it, Hedgewood or whatever, maybe they'll get sued in a big class-action lawsuit and lose a few billion dollars. A couple of bigwigs might even go to white-collar prison for a few years with Bernie Madoff and that pharmabro dude who listens to the Wu Tang Clan. But that doesn't feel like enough."

"No," Lilian said. "It doesn't. You know if Rex Garrote really was still alive, I think I'd kill him myself."

"I don't know if I could do it," Ben said. "But I want to. It's funny—"

Lilian turned to him, expecting him to continue, but he just kept walking. "What's funny?"

He shook his head. "Nothing. Never mind."

She shrugged, too tired to prod. Whatever it was, he would tell her in time. If not, it didn't matter. Survival was the only important thing now. Anything else they could deal with once they were safe inside the prison.

"I just realized I haven't gone pee for like four hours," she said, the pressure on her bladder suddenly painful.

"New world record," Ben said. "I wouldn't mind stopping by a washroom if we can find one."

Ben took out the map again and Lilian searched the legend. She found a little toilet icon nearby. They would have to follow below the tram line all the way to the creek. Ben folded the map and tucked it back into the right cargo pocket alongside the lump of his wallet and the two of them headed off toward their destination. High above them the trams hung unmoving, casting long shadows along the path.

Ben said, "Do you think Demont was telling the truth? Do you really think we'll be safe at the prison?"

"Why would he lie?"

He gave her a sharp, condescending look. "You've played enough survival horror to know that. When someone asks 'Would you kindly,' it's not always in your best interest to comply. And just because we bring him the gas doesn't mean he'll let us in."

She narrowed her eyes at him. "I trust him, Ben. He tried to help us. And anyway, this isn't a game. You can't equate everything in life to what happened in *Silent Hill*."

"*Bioshock*," Ben corrected her.

"Whatever, I was just using an example. And quoting Garrote books isn't going to help us, either."

He shrugged, watching his feet with a cowed expression. They fell into another long silence, accompanied by the sound of their shoes scuffing on fresh pavement and the occasional cry of a distant bird. The sun dipped behind a light cloud as the silence stretched out. A jet trail had cut across the clear blue sky, a stark reminder of the reality they had left behind when they'd first stepped through the entrance into Ghostland.

"Hey," Ben said, and then fell silent.

"What?"

"Were you..." He seemed to be struggling to formulate his question. "...were you really planning on ignoring me forever?"

The question surprised her. She wasn't sure how to answer it without hurting him more than she already had. She needed to be tactful, and she reminded herself the reason wasn't entirely his fault, nor her own. It all seemed so silly, avoiding him to avoid thinking about death, now that death surrounded them, everywhere they looked.

"You know my parents want me go to Stanford so I can be close to you, right?" he said.

The Stanford thing again. How did Ben know about that? "I'm *not* going to Stanford," she said.

"I told them about your scholarship."

"How does everyone know about that?"

He looked at her with a hangdog expression. "I eat lunch at the diner on Fridays."

"I knew it! You *have* been talking to my mom."

"She's worried about you. So was I."

She eyed him cautiously. "What do you mean, 'was'?"

"Well, I really don't think we'll have to worry about you anymore if we make it out of this place alive," he said. "I know you've been scared of what's out there, after high school and stuff. You're scared of taking a chance. But if you can survive this, you can do anything you put your mind to. And I'll be right beside you if you want me to be. I was thinking about listening to my parents and applying to Stanford, too."

"But I'm not *going* to Stanford."

"You say that. But I don't believe you."

He kicked a soda can in his path, as if to emphasize his point. It skittered off along the pavement ahead of them and came to a stop at the broken, bloodied corpse of a girl about their age, even dressed similarly to Lilian, except instead of ripped jeans the dead girl wore black yoga pants. Under other circumstances, it could easily have been her.

"You're gonna stay in Duck Farts the rest of your life?" Ben continued, ignoring the dead girl. "You? You're way too smart for that. You could be anything you want." He flashed a sarcastic smile at her. "I *believe* in you, Lilian Roth."

She felt her cheeks flush. "Shut up."

"I'm serious."

"*I'm* serious. Quit talking like that or I'll give you a wet willie like I used to."

Ben grinned at her. "You wouldn't."

She stuck a finger in her mouth and he held up his hands in surrender.

"Okay, okay!" He looked off. His smile faded. "There's the creek up ahead."

The pavement ended at the edge of the creek. The tram line continued over the water with a support tower on either side of the divide. Off to the left of the closest tower was a small restroom with a water fountain out front. They turned to each other, both speaking at once.

"Do you want me to wait—?"

"Should I come with—?"

They laughed anxiously. Ben held out a hand for her to finish.

"I don't think we should split up," she said.

He nodded. She could see the blush already rising on his neck. "Classic horror noob mistake," he said. "So... men's or women's?"

"Women's will have more stalls," she said. "For privacy."

"You just want to trick me into going into the girl's room, like that time you pushed me in."

She laughed. "It was way more than once." She used to love pranking him when they were kids because he was so easy to embarrass. Thinking back on it, it had been pretty mean, considering how kids bullied him later in life.

Ben said, "I've blocked the rest out of my memory. Sandy Lewis was in there taking a dump that one time, she stank the whole place up. No one needs to smell that."

She laughed again. "Hey, girls poop, Ben. Get over it."

He laughed with her, then his expression got gravely serious. "Wait, you're not...?" He wouldn't finish the question but she'd already heard enough.

"Don't be gross."

"Girls poop, Lilian. But if you do have to, I'm waiting outside."

"I've got a wet finger with your name on it, Laramie...."

"Good times," he said with a chuckle. "Let's get this over with before we piss our pants."

Lilian entered the washroom first. She held the door for Ben. Three stalls stood directly across from the long mirror and counter, sinks with automatic faucets and soap dispensers. Despite being its first day of use there was already water splashed on the counters, soap drooling from the dispensers, smudges on the mirror, the trash can overflowing with crumpled paper towels and a few scraps of toilet paper scattered on the floor.

Lilian headed straight for the last stall on the left. Ben pushed on the door closest to the entrance, obviously hoping for some buffer between them. The stall was locked. Lilian glanced at the mirror but she couldn't see any feet under the door in the reflection.

She closed the door behind her and twisted the lock. Ben entered the middle stall as she unzipped her fly, wriggled her jeans down over her hips and sat down on the cold toilet seat. She heard him unroll a bunch of toilet paper and unbuckle his belt. She relaxed and began to pee, trying not to think about Ben in the next stall hearing her go. As she did, she glanced down at her feet, tapping anxiously, and saw two fresh spots of blood in her underwear.

Great. I'm not supposed to get my period for ten da—

A drop of blood landed on the stubby, bitten nail of her index finger. As she twisted her arm to get a better look, another drop struck her forearm and spilled down, mingling with the light brown hairs.

Reluctantly, not wanting to discover the source of the blood with her pants still around her ankles, Lilian looked up.

She sucked in a gasp.

The woman from Guest Services floated above the stalls. She still wore her Ghostland T-shirt over her heavy bosom and her hair hung stringy from either side of her head. She looked frightened and confused and she was weeping softly, a muttered whisper of prayer, holding out her hands to Lilian, the splayed fingers covered in blood from two gaping wounds she'd slashed in her own wrists, the razor blade still pinched between thumb and forefinger, dripping blood onto Lilian's hair and jacket and face.

Lilian screamed and jerked up her pants, pushing on the door until she remembered it opened inward, then fumbling with the latch, still pulling up her underwear with the other hand.

Ben hurried out of the stall beside her. He glanced at her crotch, with her pink underwear still clearly visible above her unzipped jeans and then followed her haunted gaze to the ceiling.

He saw the woman and staggered back against the counter. "Jesus Christ!"

At the sound of his voice the dead woman from Guest Services spun around in midair above the stalls and floated toward them, still mumbling, still holding out her slashed and bleeding wrists. Lilian turned, looking behind herself, and somehow it was the fact that the dead woman didn't cast a reflection that shocked her into action. She grabbed Ben's sleeve and pulled him toward the door, almost tripping with her jeans still below her hips as the two of them ran screaming like frightened little kids who'd just summoned Bloody Mary in the school bathroom.

"That was close," Ben said.

Lilian sighed. Ben watched her wriggle her jeans up over her hips and turned away shyly as she forced up the zipper.

"You think she'll come after us?" he asked, after a moment of watching his shoes as he ground dirt into the pavement.

"I don't know. I think she's just scared and confused."

"Until Garrote gets into her," Ben said. "Then she'll be just as dangerous as the rest of them."

Lilian went to the fountain, pressed the lever with her elbow and ran her hands under the cold stream. The dead woman's blood had already vanished—it had either been a hologram or a hallucination—but it still made her feel disgusting. Once she was satisfied her hands were clean, she dried them on her jeans and drank a few mouthfuls. She hadn't realized how thirsty she'd been until the cool liquid splashed the back of her throat. It felt glorious.

Ben approached the fountain. He drank greedily and let out a satisfied gasp.

"You're not gonna wash your hands?"

"I didn't touch anything."

"You touched the doors."

Ben groaned and rolled his eyes. "*Fine.*" He rinsed his hands briefly and wiped them on his pants. "You happy?"

"Better. Just because it's the end of the world doesn't make hygiene any less important."

Ben shook his head and chuckled. Lilian chose to ignore it. They headed off from the washrooms in a westerly direction, following the creek until it branched north.

A little way further the Museum of Haunted Vehicles stood at the end of a wide courtyard, a flat glass and chrome building that looked a lot like a brand-new car dealership. Lilian thought it must have been built specifically for Ghostland, unlike other buildings that had been shipped here in pieces or like Garrote House, carried on the back of a truck. Aside from a dozen or so scattered bodies, sprawled in various painful positions, an upturned Ghostland-red golf cart and several park benches, the courtyard appeared to be empty. Flies buzzed on and around the corpses but the area smelled much fresher than elsewhere. The wide-open space must have allowed the spring breeze to circulate the air.

The two of them headed out across the courtyard. It was so quiet they could hear their footfalls echoing back to them from the building.

As they approached the steps, a large figure bolted out from the hedge maze to the left of the courtyard in a crouch. Lilian recognized the big guy from the front gates, the one with the gun permit and the NO FEAR t-shirt. The man seemed to have forgotten the motto as he crouch-ran into the middle of the courtyard, whimpering and shading his eyes to look up at the sky.

"Why is he—?" Ben began.

No Fear let out a blood-curdling shriek, silencing Ben as a dark shadow shot diagonally across the courtyard toward him.

Without a word, Ben and Lilian dashed up the stairs. Lilian jerked open a door and slipped inside, holding it until Ben got through, then pushing it shut on its pneumatic hinges behind them.

"Holy hell," she gasped, leaning back against the door. "What the hell are those things?"

"I don't know," Ben said, sounding utterly winded.

No Fear stood and ran for the stairs, screaming and panting. Three more dark shapes fell from the sky, looking something like billowing black cloaks or jellyfish made of smoke. One completely enveloped the man. Another missed him and circled back. The third struck the ground and splatted on the concrete, leaving a wet black stain.

No Fear's terrified scream ended abruptly, like a needle coming off a record.

A moment later the smoke creature rose into the sky, leaving the corpse on the pavement. Its ethereal limbs fluttered as it rose to join the others, as if it was swimming instead of flying. The creatures circled like buzzards looking for prey, and then they melded into one another, forming a nattering, swirling mass of death.

"*The Swarm*," Ben said, and shivered.

GHOSTS IN THE MACHINES

"IS HE DEAD?" Ben whispered. The two of them were crouched behind the museum directory, peering out through the glass front of the building.

"He's not moving," Lilian said. "What are they doing out there?"

The Swarm had broken apart into individual shadow creatures, most of which were circling above the building. The others—and there had to be a dozen of them, maybe more—swam languidly through the air toward the building's glass front. Apart from the Swarm they appeared to be eyeless, completely devoid of features aside from their wispy tendrils. And yet they seemed to be looking through the glass wall, searching for the living beings that had escaped them.

Were they demons? Ghosts? Some other sort of dark presence? Ben wasn't sure. He supposed the Swarm could have been a literal version of what Sara Jane Amblin had mentioned: an oil spill, a massive pool of raw dead energy stripped to its essence, devoid of any semblance of its former animal state. A pure and lethal entity whose only intent was death, with a one-inch thick sheet of glass separating them from their next meal.

"I dunno," Ben replied. "Maybe they can't get through."

As he said it, he noticed large black smudges on several of the windows further down, where the soul-suckers had likely launched themselves at the windows, rocketing after intended victims, vanishing as they struck the glass the way the one outside had into the pavement. The fact that these creatures weren't eager to do the same meant they'd learned from the mistakes of the others. Intelligent hunters with a hive mind.

It also meant at least one person had fled inside. They could still be in here, for all Ben knew.

"Ben...?"

Lilian was peering further into the building when he turned, breathing heavily with her eyes wide.

Three vehicles rolled toward them from between the exhibit stands, the buzz and squeak of their small engines and parts growing louder as they approached. In the lead was a slightly smaller than normal tricycle, push-pedaled by a doll[1] in a dusty white Victorian dress, with long black hair, a jagged crack down its pale face through the missing left eye and a black bowler hat on a slight angle. Behind it was a cherry red car that looked like the car from *Christine,* and a dune buggy with an old G.I. Joe doll buckled into the driver's seat, the large ones that looked similar to Barbies. All three came to a stop a few feet from the end of the aisle, then buzzed forward and back like kids on bikes trying to intimidate them. Ben guessed they were possessed by poltergeists too uncreative or weak to manipulate something larger and more menacing.

Before he had a chance to avoid it, Christine buzzed toward him and struck his ankle. "Yow!" He raised his foot off the floor. The red muscle car idled at his feet. There were stickers plastered all over it. Some kid had gotten a lot of enjoyment out of it. Ben wondered if that same kid's ghost possessed it now.

"Little bastard," he said. He kicked at it but the car was too quick, darting back out of the way of his shoe. The other two rolled around to either side, the tiny engine buzzing, trike wheels squeaking as they flanked him and Lilian.

The dune buggy whizzed up to her. She jumped and it zipped under her feet. "Hey! What gives?"

The doll on the tricycle cocked its head to look up at them, staring with its one remaining blue eye. "*Wanna ride with me?*" it said in a hollow recorded voice. "*C'mon! Let's be friends!*"

"Get lost, you creepy little shit," Lilian muttered, edging past it.

Ben stepped over the red car. It didn't move to follow him.

As they headed down the aisle into the exhibits, the two cars zipped off in different directions while the doll rolled around in a wide circle to face them, its wheels making tiny birdlike squeaks. The doll raised a hand from the handles and waved side to side. Ben noticed several fingers had chipped off. Its right knee rose and fell, then its left, and it rolled slowly off to the right until it disappeared behind the exhibit stand at the end of the aisle.

Outside the dark entities hovered, patiently awaiting the kill.

Ben forced himself to look away and not turn back, like a child hurrying up the basement steps, certain he'd seen the bogeyman.

The building wasn't as large as he'd expected—there appeared to be about twenty exhibits, maybe a few more. Automobiles from different eras stood on platforms to the left and right. A mint-condition Rolls Royce Silver Ghost. A beaten-up yellow VW Beetle. A bright-red hotrod with flame decals on the sides. An antique elevator car on a stand. A large unicycle behind a plexiglass case. As they cautiously headed down the aisle, an '80s rock song he almost recognized drifted from somewhere within. It sounded like music his parents would play in their room when he was supposed to be asleep. That was reason enough for it to give him the creeps.

Each exhibit bore a sign with its previous owner's name and date of death, like the ones found elsewhere in the park, along with a corny pun tagline and its history printed below. Each sign also featured a button for an audio version. Ben didn't care to stop and check them out but he and Lilian cautiously peered in through the windshields as they passed to be sure no one was inside and to check their gas gauges. Each vehicle sat on Empty.

On the other side of the intersection, a mobile home, an old train engine that smelled of burnt coal, a subway car covered in graffiti—the sign called it CAR 438[2], and Ben knew some of the New York subway car's haunted history from an episode of *Ghost Brothers*—a houseboat and a white van all stood on raised platforms. Over the roof of the trailer Ben could see the dirty shovel of a massive excavator and it offered him a brief glimmer of hope knowing they could probably tear down the whole damn exterior wall of the park if they wanted to, until he remembered neither of them knew how to drive an excavator. He doubted it would be easy to learn.

Christine zipped down the end of the intersecting aisle to their left, following them. He looked back, expecting to see the doll on the trike had creeped up behind them. But all he saw were the shadow creatures hovering behind the glass, linking tendrils, still waiting for their meal to return.

THE ROCK MUSIC was louder among the marine vehicles. Ben recognized the song from its repeated refrain: "Rosanna," by some '80s band he didn't know. It came from a yacht down the aisle to their right. There was movement up on the deck but from the low angle he couldn't see what it was. Closer to them were the mangled remains of a silver sports car—the exhibit labeled "*'Little Bastard,' James Dean's cursed Porsche 550 Spyder!*"—and an Econoline van featuring the vanity plate FILTHY and a bumper sticker with the words F*CK GONZALEZ, whoever that was. Ben cupped his hands to peer in through the tinted windows. The orange gas gauge needle floated just above the middle.

"This one's got gas," he said.

Lilian was looking up at the yacht, watching the ghosts on the deck. Painted in light blue sea spray on its stern was its name: *Sea Dream*[3]. A gaping hole had been torn in the hull with barnacles encrusted around it. On the deck, a blonde woman in red plastic sunglasses and a high-waisted yellow bikini lay tanning. A tanned blond man in a white Polo shirt, high-hemmed yellow Adidas shorts and tennis shoes stood at the wheel pouring champagne. The man raised his glass in a cheer. As he turned to drink, the barnacle-encrusted skull on the other half of his face came into view, like a mass of gray and black tumors. Frothy liquid spilled out through his exposed teeth. His partner raised her glass and a slippery green eel oozed out from between her cherry-red lips and plunged its snout into the glass, guzzling the golden fluid.

Ben flinched away in disgust. He flipped open the Econoline's gas cap, urging himself to stay focused on the task at hand and not look back at the ghosts onboard the *Sea Dream*.

"Great," Lilian said, watching him work. "Now we just need a hose and a gas tank. Unless you plan on holding it in your mouth."

"My mouth's not big enough," he said. "Yours might work, though."

"Ha ha."

Ben crept around to the back, keeping an eye on the ghost yuppies on their yacht, still lost in a world of their own, drinking champagne while everything around them had fallen into chaos. The back doors wouldn't budge and the window was blacked out. He heard a click and saw Lilian gently open the passenger door. She crawled into the seat. "You gotta be kidding me," she whispered.

"What?"

He heard a thump from inside the van. Lilian had disappeared from the windows.

"Lil? Are you okay in there?"

A moment later she backed out of the same door she'd entered. "How lucky are we?" she said, holding up a red plastic gas can and a length of hose.

Ben rolled his eyes. "Yeah, sure. Super lucky."

"Are you gonna take them?"

"Why me?"

"I've never syphoned gas before."

"Neither have I."

"Well, it's easy," she said. "All you do is stick one end of the hose in the tank and suck on the other end until gas starts coming out, then you stick it in this thing." She shook the empty can.

"If it's so easy, why don't you do it?"

"If I do it, I'll puke. I hate the smell of gasoline. I've already puked once today, it's your turn."

"Nobody likes the smell of gas, Lilian."

"What about Skylar Peterson?"

"Do you seriously think she's sniffing gas because it smells good?"

Lilian considered it and shrugged. "Come on, Ben. Just do it."

"Fine. But you owe me."

"Great. If we make it out of here alive, I'll set you up with one of my friends."

"Oh, please let it be Skylar," he said with heavy sarcasm. He took the can and hose, sat the can beside the rear wheel and unscrewed the top. He removed the gas cap, the pungent smell sharp in his nostrils as he shoved a length of hose down inside. He hesitated with the end of the hose poised before his open mouth.

This better be worth it, he thought.

"Don't swallow it," Lilian warned him.

"I wasn't planning to." He sucked on the hose, feeling the pressure build, like sucking a thick milkshake through a thin straw. The gas fumes started to make him feel dizzy. He sucked harder and a mouthful of awful-tasting fluid filled his mouth. He spat it out. Gas poured out of the end of the hose, splashing on the floor at his feet. He jammed it quickly into the gas can and watched the can grow darker red from the bottom up as the gas began to fill it.

"Good work," Lilian said.

Ben worked up a mouthful of gas-flavored spit and hocked it onto the platform. "No thanks to you," he said, though he had to admit he was glad it had washed away any residual taste of the nun's spit from his mouth.

Somewhere nearby an engine sputtered to life. Lilian looked off in the direction of the sound. "Sounds like someone's got the same idea as us."

The sound of the engine made Ben nervous. He looked down at the can. It had already filled to the quarter mark but he willed the can to fill faster. How long would half a tank last them? A few hours? The park wouldn't close until nine but late arrivals after work might wonder why the entrance was barred and word would eventually spread to the local police. He guessed it had to be at least three or four o'clock now, maybe even later. It felt like days since they'd left the funhouse and struck out for Guest Services. So much had happened since then. So many people had died.

"I smell cigar smoke," Lilian said.

Ben sniffed and caught a faint whiff over the heady fumes.

He remembered the gangster ghost in the flat cap, firing this Tommy Gun with a cigar stub in his mouth, and hoped he was wrong.

The tank had filled about three-quarters of the way. It would last several hours, maybe even all night depending on what Demont was powering. "Good enough," he said. He screwed on the cap and they continued onward, eyeing the deck of the *Sea Dream* as they slipped between it and the run-down mobile home.

"Psst! *Psst!* Hey, kids!"

The man was sitting on the porch steps of the trailer home with his elbows on his knees. Ben thought he looked vaguely familiar: handsome, clean-cut, short dark hair. He wore a loose blue chambray shirt with the sleeves rolled up, sweat-soaked at the underarms and tucked into his khakis. A brown leather belt matched his shoes; all three had shiny brass buckles.

Was he a ghost? Ben had no idea, but it didn't seem like he meant them harm. He looked scared himself. The trailer looked in worse shape. Dingy curtains hung like a dead woman's limp hair behind dusty, cracked windows and screens with holes. The aluminum siding, once white, now almost purely gray with age and grime, had rusted and peeled back in places, leaving jagged, dangerous edges. The roof had collapsed on one side. A pile of dead leaves lay against the inside of the screen door.

"Little help?" the man asked.

Ben took a step forward. Lilian grabbed his shoulder, pulling him back.

"That's him," she whispered.

"Him who?"

Ben saw the handcuffs securing the man to the rusty, paint-flecked railing. He'd been trying his best to hide them behind his hip but there wasn't much room on the small staircase leading up to the trailer home. And then Ben realized where he recognized the man from: it was Detective Beadle's suspect, the rich kid who'd gotten away with murder. Stan must have confronted him here, managed to cuff him and then went off to find help before the whole damned place went to hell.

Stan might still be alive, he thought ecstatically. But he wouldn't get his hopes up. There'd already been enough disappointment today.

"What's with the handcuffs?" he asked the man.

The murderer chuckled. He rattled them with an awkward smile. "Someone's sick idea of a joke. Listen, I need help. It's not safe for me here." He glanced back at the screen door. It banged lightly against the jam, twice, as if someone with very little strength was trying to open it from the inside.

"We know who you are," Lilian said, hands on her hips. Ben knew that pose. It meant the man didn't stand a chance.

"What do you mean, dollface?"

"Don't call me that. You're Alex Fischer. You're the Doll's Head Murderer."

The man chuckled again, this time with an edge of fear that rose to frantic within a split second. "That wasn't me, okay? That stupid fucking *pig* tried to railroad me! You gotta help me out here, please. I am *begging* you here!" He clasped his hands together, the cuffs jingling as he shook them. His perfect coif shook loose over his sweat-drenched forehead. "Please, *God*! It's *not safe* here! Please, you *have to help me*!"

"Would you shut up a second!" Ben shouted. The man stopped screaming and began to fidget with the cuffs. Ben turned to Lilian. "What are we gonna do?"

She shrugged. "We can't let him go. He's a murderer."

"Suspected murderer. Okay, he probably did it. But we can't just leave him here to die."

"What was he doing here if he didn't kill her then? Paying his respects? You know he did it, Ben."

Ben didn't believe in capital punishment, an eye for an eye. Murderer or not, he couldn't just let the man die. It would only leave more dead energy for Garrote to absorb, and if Detective Beadle was still alive out there, Fischer's ghost would hunt him down and torment him. He likely wouldn't let the two of them get away with abandoning him here, either.

"What happened to Stan?" Ben asked.

"Who?" Fischer said irritably.

"The detective. Where is he?"

"How the fuck should I know? He just left me here. Kept saying he hoped Valencia would get her revenge, I don't even know who the fuck Valencia is, I just came to see the ghosts! I *paid* to see the *ghosts*!"

A tricycle bell rang—*ring-ring!*—and the doll rolled out from behind the trailer. The two remote controlled cars had stopped at the opposite end, the three small vehicles blocking them in.

"*Oh goodie! More friends!*" the tricycle doll said.

Ben heard the sound of the engine again, closer now, before the killer started shouting, kicking out at the cars and trike as they rolled up to the porch at his feet. "Get the fuck outta here, you little shits! Go on, get lost!"

His foot struck the doll in the head and the trike tipped over with a rattle of the bell. "*Not fair!*" the doll cried. "*I'm telling mommy!*"

A shadow fell over the doll and its trike. In the same instant the engine roared, close enough to make Ben jump. The bullet-riddled Model T rolled into the aisle, its engine wheezing like a dying fan as it wheeled around on its whitewall tires until the headlamps illuminated Fischer like a prison-yard spotlight. In the darkened driver's seat, the red glow of a lit cigar illuminated the face and flat cap of the driver. A ghostly puff of smoke rose from the opened window.

Fischer spat at the car. "Fuck you! You don't scare me!"

Maybe now he doesn't, Ben thought. But he'd seen the driver's Tommy Gun and he didn't want to find out what sort of damage these phantom bullets could do. He grabbed Lilian by the arm and the two of them backed away quietly.

"Run," he said in a near-whisper.

"What?"

"*Run*," he shouted, and he turned and ran himself.

AS THEY RAN Lilian heard the murderer screaming for his victim to save him, calling her *sweetheart* and *bitch* in the same breath, until finally his screams rose to a tremulous soprano and the wheels skidded and the engine revved and with a jangle of metal and glass the killer's cries suddenly stopped.

Skidding on the soles of her shoes, Lilian banged her shoulder against the side mirror of a beat-up red pickup truck. The exhibit sign read like a bad pun in a newspaper headline: *Stunt Driver Loses Head, Job*[4]. Lilian tugged on the passenger door. It opened with a squall of rusty metal. She rose on her tiptoes to peer inside, then waved Ben over.

He dashed across the aisle, leaping over Christine as Lilian slipped inside. He chucked the gas can in the truck bed and slipped into the passenger seat. The interior smelled like cigarette smoke and beer and a strong, woody cologne. A pair of fuzzy dice hung from the rearview mirror and the driver's visor was down, photos of naked women cut from magazines taped to the inside. The keys were in the ignition. The gas needle sat on ¼.

"Oh, heck yeah," Ben said. "Can you drive this thing?"

Crouched beside the steering wheel, out of sight through the window, Lilian shook her head.

"I thought you got your learner's permit that summer you spent in South Dakota with your cousins."

"That was four years ago," she said.

"Lilian. You're the only one who knows how. My parents would never let me drive, you know that."

"I'm afraid to drive, okay? Jeez!"

"You can do it, Lilian. I believe in you."

She shook her head.

"Would you rather get shot?" He poked his head above the dashboard and peered out the windshield. "Look, it's just a straight shot past the excavator to the doors. I'll steer, all you gotta do is step on the gas."

"If it's so easy, why don't you do it?"

"Because there's three pedals and I don't know which is which!"

She heard a tiny horn honk just below the driver door. She couldn't see the toys or the car—the shooting had stopped shortly after they'd hidden in the truck—but she heard the large engine roar in response to the honk, and the trike bell's subsequent *ring-ring*.

The ghosts were working together. If she didn't get them out of here soon Ben was right, they were dead.

He turned the key in the ignition. The engine sputtered.

"You have to press the clutch," she shouted at him.

"Which one's the clutch?"

"The left one! The left!"

The truck lurched as Ben raised off the gas and pressed down hard on the clutch while twisting the key. The engine rumbled to life and the stereo came on full blast, some woman singing about a "magic man." Lilian threw the transmission into first gear. The front wheels thudded down off the platform and the truck immediately began to veer to the right. She grabbed the steering wheel and tried to hold it straight.

"Ram the doors!" He was pressing the pedal right down to the mat.

She glanced in the rearview just as the gangster's car slammed into the platform behind them. The ghost leaned out the window with his machinegun, the cigar clenched between his teeth, and aimed down the barrel.

"Floor it!" she cried.

Ben hit the gas.

THE TWO OF them ducked below the dashboard as phantom bullets struck the tailgate and shattered the back window, a shower of glass falling around them, pattering on the seat fabric. It wouldn't be long before the Model T caught up to them and he filled them full of lead.

"Ben!" Lilian shouted.

"What?"

"*Bennn!*"

He rose just enough to peer over the dash and see the excavator shovel lowering down on them. They passed below it and for a moment he thought they'd made it through unscathed, but the big metal teeth screeched along the roof and slammed down into the truck bed.

"The gas!" He jumped up in his seat, looking out the back. The shovel had tipped the gas can and scraped along the plastic bed cover. Luck blessed them and the tailgate tore open as the shovel struck it and smashed down on the concrete floor behind them, shaking the ground as the truck kept barreling forward.

Ben kept his foot pressed firm on the gas while Lilian took her right hand off the wheel to jerk the transmission. There was a grinding sound. It sounded bad, like trouble. He saw her move her right foot to the center pedal and with her right hand still on the transmission the truck began veering off, toward the thick cement pillar between the doors and the glass wall, as she finally rammed the transmission to second gear.

"Hold it straight!"

"I'm trying! Stupid stick shift!"

The truck missed the pillar by an inch on Ben's side and struck the doors dead center. Shattered glass pattered on the roof as the exit doors flung outward and the truck bounded down the steps into the courtyard. Bodies littered the ground ahead of them. Lilian navigated to avoid them. Ben remembered the shadow creatures and looked up into the sky just as a black shape fluttered over the sun.

"Lilian!" he cried, raising his arms feebly to protect himself.

She saw it and screamed, jerking the wheel to the right. Several shadows hit the truck, *thud-splat thud-splat* like a rain of toads, one after another. She kept her foot on the gas, still ducking, expecting the creatures to break through the windshield and suck their souls out of their bodies.

Nothing happened.

The truck lurched over a bump. He heard the gears grind again as she jerked the stick to the 3 and then the windshield wipers came on, squeaking across the remaining glass.

"You can look now," she said.

Ben sat up straight. Lilian had driven them to the other side of the courtyard. The wipers streaked away black ectoplasm, clearing the windshield of debris.

He sighed in relief. "You did it."

She turned to him and smiled. "I'm not afraid of driving anymore."

"That's great," he said. He meant it.

She stepped on the gas. The truck bounced as she ran over a dead body on the ground. "Sorry!" she called over her shoulder, as if the corpse could have heard her. As if the corpse *cared*.

Ben looked out the back to check on the gas can. It was still there, lying in a pile of broken glass. A trail of glistening liquid had run down the corrugated bed cover. He hadn't managed to get the lid on at the right angle—plastic screw tops were the worst—and he supposed some must have spilled out when it fell down. He brushed away glass from the ledge, leaned out and stood the can up against the back of the cab.

When he slipped back into his seat Lilian was smiling to herself as she turned the truck out into the promenade. She turned up the tape player and flicked the visor up like they hadn't just escaped a serial killer and a pack of possessed vehicles intent on killing them, and they were just going for a cruise through the streets of Duck Falls.

Ben shook his head and laughed.

SANCTUARY

BEN KEPT QUIET as Lilian rolled the truck up to the prison gates, biting back the urge to beg her to turn around and drive all the way back to the front gate. For as dangerous as the idea sounded, without the code for the service hatch their only option was to hide out and wait for rescue. If Demont's fortification worked as well as he claimed, it was entirely possible Fontaine County Correctional[1], the most haunted prison in America, was also the safest place to hole up until the cavalry arrived. If they could rally more survivors—if there was anyone left to rally—maybe they could salvage something from this tragedy. Turn a brutal tale of loss and desperation—a story Garrote himself could have written—into a tale of grit and survival against all odds.

"Are you still with me?" Lilian said.

The question took him off guard. "What? Of course, I'm with you, Lilian. We're a team. It's you and me or nothing."

She smiled. "Good. Because we need to watch each other's backs. We need to keep each other alive."

"You can't get rid of me that easy. If I die, I'd probably still haunt your ass."

"That's just what I need," she said. "A haunted ass."

Lilian laughed. He couldn't help but join her. Then she nodded very seriously and took her foot off the brake. She stopped the truck just shy of the high wrought-iron gate. Looking up at the wide-open doors, he wondered how the prison could possibly be secure. Ghosts could pass through walls. Bars and locks wouldn't deter them. But he was too tired to fight, too beaten down. He needed to rest. Instead, he flicked on the walkie.

"Demont, this is Ben Laramie. Over."

While waiting for Demont's reply he stared in the side mirror, watching the midway, searching for movement. A few bodies lay in the street. One of them was a girl just a little younger than the two of them, probably thirteen or fourteen, hugging a giant bright yellow *Minions* doll as if she'd thought it would protect her from the monsters. Nearby, the man from the waffle stand hung halfway out the service window, his arms outstretched, eternally reaching for his white paper hat on the pavement below.

Ben glanced out the windshield but his gaze kept returning to the paper hat. He kept waiting for it to blow away, further down the midway.

Demont's sudden reply startled him. "*Hey, sorry, had a bit of a problem I had to take care of. Let me know when you're at the door and I'll open it for you.*"

Lilian turned to Ben as if for final confirmation and even though his every nerve fought against it, he gave her a sharp nod. She pressed on the gas and they passed under the stone arch.

Ben turned to look out the back. All of his many years playing video games, reading and watching horror had prepared him to expect the worst. He expected to see a horde of shambling zombies rushing the gates, alerted by the rumble of the truck. But the street behind them was empty. In the midway, the waffle man's hat finally caught in a breeze and fluttered away, rolling off into the distance.

"Okay, we're through," he said into the walkie.

He turned back to the prison. Lilian thought sanctuary lay behind its walls but when Ben looked up at its impassive stone face he felt like a condemned man on his way to the electric chair. He'd been extremely lucky once, surviving his heart attack. He didn't think it was likely he would be that lucky again.

All he wanted was to see his mom and dad one more time.

No, forget that. This close to death, for-real death, was no time to be reluctant. He wanted more. He wanted to kiss a girl, go to college, get a job working at some online newspaper and rent an apartment in the city. Get married, have kids, change diapers, grow old. All the stupid sappy shit his parents and their friends talked about when they got together that made him zone out and retreat into a fantasy world of ghosts and monsters, he wanted all of that and more. So he promised himself they weren't going to die here today. He wouldn't let it happen. He'd fought too hard and too long—not just at Ghostland but every day since Garrote House had rolled through town, almost killing him the first time—to give up now.

No half-assing it, he thought. *Gotta play in Hard Mode now.*

"Where's the stun gun?" he asked.

Driving slowly toward the prison, Lilian rose slightly from the seat and nodded toward her back pocket, where the electrodes poked out above the seam. Ben caught it gingerly between thumb and forefinger, trying his best not to touch her butt while tugging the weapon out.

The main building was a large sandstone rotunda with an impossibly high ceiling of tinted green glass. Two wings swept back in either direction, creating what looked like a giant bird in the overhead view. According to what he'd read about the place, it had been the last of the old maximum security "roundhouse" prisons. Every cell in the four stories of its main block was visible from the central guard tower. Inmates had called this "the Circle of Death," since many of them had been facing life sentences. Some called it "the Colosseum" due to its shape but also because brawls had allegedly been allowed to go on long enough for one or another prisoner to die, while the guards watched from the safety of the tower like Roman patricians, betting on the outcome.

Lilian stopped in front of the battered blue metal front door and unbuckled her seatbelt. "Ready?"

He nodded, unbuckling himself. They climbed out of the truck and looked around. The expanse between the front gate and the prison was deserted but he suspected it wouldn't remain empty for long. If there were any infected ghosts within shouting distance, at some point very soon, Garrote would find them.

He sees through their eyes, Ben thought, and shuddered. He thumbed the Talk button on the walkie. "We're here," he said, peering over the back of the truck into the bed. The corrugated black plastic glistened in the late-afternoon sun. He traced the spill back to a large gouge in the red plastic container, dirt around the edges of the hole. The excavator must have done it when the shovel dropped down on the truck. "Shit!"

"What's wrong?"

"There's a fucking hole in the gas can."

"Oh no, Ben..." She shook her head in disappointment.

He picked it up at an angle so no more gas would spill out and sloshed it around to gauge how much was left. "Not much in it. Less than a quarter of what we had."

"After all that?" She scowled at the prison door. "Where the hell *is* he?" Just as she made to pound on it, the door unlocked with a buzz.

"*Door's open,*" Demont said over the walkie. "*Come on in.*"

Lilian looked back over her shoulder. Ben shrugged. She pulled the door open and stuck her head in. "He's not here," she said, her voice sounding hollow.

"*Glad you made it,*" Demont said over the walkie. "*We're running on fumes in here.*"

"Where are you?" Ben asked.

"*Up in the control tower.*"

"Aren't you gonna come down?" Lilian asked.

"*For your safety, it's best if I stay up here. I'm monitoring you on the security cameras. But you guys better get your butts in here quick, I see some nasties heading toward the gate.*"

Ben shot a cautious glance behind them. He didn't see anything out there but he wasn't about to wait around for Garrote and his "nasties" to show up. He followed Lilian into the cool, dank, dimly-lit building. Several dead lay on the floor and against the walls, clearly in the process of fleeing when the park went into full meltdown. A generator rumbled somewhere inside. Likely beneath the rotunda, judging by the echo. Below it he could hear a vague vibrating hum he could feel in his fillings.

Electricity, he thought. Lots of it.

Lilian already stood just beyond the administration desk near the barred door to the prison interior, tapping a foot impatiently with her arms across her chest. She'd slipped her jean jacket back on against the chill. Ben hurried up to her and reached for the barred door leading to the cell block.

"*Don't touch the bars,*" Demont said. "*Not unless you wanna get knocked out of your shoes.*"

Ben drew his hand back quickly, spotting a copper wire curled over the bars. It ran between the two doors where the floor met the wall to the inner door, wrapped around the bars there and spooled into the dimly lit cell block.

"You electrified the bars," he said.

"*Good eye. You see that broom handle over there by the desk?*"

Ben saw it leaning against the wall.

"I got this," Lilian said, grabbing it in both hands like a weapon.

"*Easy now. Use it to flick the copper wire off the bars.*"

Lilian did, as cautiously as she could manage. The hum Ben had heard since they'd entered the prison diminished, now coming only from beyond the inner door and inside the rotunda.

"*Great,*" Demont said. "*Now come on through, the door's unlocked.*"

Lilian stepped into the entryway. Ben followed her, the gas can sloshing against his hip. He pulled the door shut behind them.

"*Lilian, do you see the bent nail on the end of the broom handle? See if you can hook the wire back up onto the bars.*"

Lilian tried a few times to snag the wire with the nail. It caught on the third attempt and she raised it delicately back to the closed door and lowered it gently onto the crossbar. Immediately the hum grew louder. Ben was close enough the hairs on his arms stood on end.

"*There was a ton of copper wire just lying around in C Block. I figure they must've stored it there to use elsewhere and never got around to it,*" Demont said.

"That was lucky," Ben said.

"*I don't believe in luck. Now this next one's a little tougher. You have no idea how hard it was to do it with just two hands. You ever play* Operation*?*"

"Once when we were little," Lilian said with a sadistic grin. "Ben told me all the kids do it."

He felt his cheeks burn as she laughed. The story wasn't true but Demont laughed along with her.

"*Way to go, Ben! But you're thinking of 'doctor.' Anyway, just keep the wire from touching the bars. And Ben, you open the door really carefully while she does. Soon as he's got the door open far enough you can let go of that wire but until then keep it steady or he'll get zapped, you got it?*"

"Gotcha," Lilian said. She slipped the broom under the wire and lifted it expertly off the crossbar. The hum diminished again.

"*Careful now. Keep it steady.*"

Ben reached for the handle, eyeing the wire. A small movement in any direction would put it in direct contact with the metal bars and he'd be toast. "Please don't kill me," he said to her.

"I won't. I don't need you haunting my ass."

He wiped sweat from his eyes and grabbed the handle, twisted it and pulled the door open carefully. Lilian matched his movement with the pole, keeping the wire from striking any of the bars. But her arms quickly started to quiver from the strain.

"I think we can get through," he said.

"Good. My arms are killing me."

She let the wire drop back onto the bar. The hum filled in the anxious silence as they stepped through, narrowly avoiding the wire.

"*Okay, now just tug the door closed with the pole and we're all good.*"

Lilian pulled the door shut, its loud clang echoing throughout the massive circular cell block. She sighed heavily and laid the pole back against the wall.

"*Nice work,*" Demont said over the walkie.

They turned to look at their electrified fortress. At the center of the Colosseum stood a guard tower. Three flights of metal grate stairs led up to an octagonal room about twenty feet from the scuffed cement floor, enclosed within tinted protective glass. From the four levels of cells surrounding the so-called Circle of Death—there had to be at least fifty cells per floor, which at two prisoners per cell would make four-hundred at capacity, although there wasn't likely to be even half as many ghosts, considering the total park tally had been just over three hundred—it would have been impossible for any of them to see into the tower, impossible to know when guards had been watching them.

And at the moment, the cell block lived up to its name: like anywhere else in the park, the dead lay where they'd expired—it reminded Ben fleetingly of the game they used to play in kindergarten, when the teacher said "ashes, ashes, we all fall *down*," and all the kids would drop where they stood.

It was clear Demont had kept himself busy while the park went to hell. He'd run copper wire around the bars of each cell on the first floor and up the stairs. Ben could see the wire trailing in and out of several cells on the second floor. He couldn't tell if the other floors had been wired but Demont had obviously had his hands full while they were out there trying to stay alive.

The generator was larger than any he'd seen outside of a construction site, and orange cones had been placed around it. In here its rumble was so loud he could barely hear Demont talking over the walkie. His mother probably would have said the racket could "wake the dead."

"Heyo, can you see me waving?"

Demont stood at the top of the stairs, leaning out the door, waving down at them. The sun was so bright through the glass ceiling above it was almost as though it shone right through him.

"Yep, we can see you," Ben said, waving back.

"All right, just get that gas in the generator and get your butts up here."

DEMONT OPENED the door for them as they reached the top of the three flights of stairs. He smiled and stepped back, allowing them to enter. "Welcome to the fishbowl," he said.

A hexagonal off-white desk stretched around the perimeter. The rolling chairs had orange tweed fabric and burnished brass legs. Heavy green sea glass ashtrays lay in front of several chairs on the desk where old TV monitors had been mounted at intervals. The prison had been closed down in the mid-'90s after a door malfunction had instigated a riot which had ended in the deaths of seven inmates, three guards and the overnight physician. Ben assumed they had retrofitted the prison with new security monitors and cameras in the time between, and that Ghostland must have put the old stuff back in to give it a 1970s vibe. Either way, none of the monitors were on at the moment. Not that they were necessary: the entire Colosseum was visible through the reinforced glass, as advertised.

"Is there anyone else here?" Lilian asked.

Demont scowled, closing the door behind them. "You mean ghosts or people?" Ben realized he wasn't wearing glasses. They were on the desk beside a folded pile of shiny silver fabric.

"Either."

"I haven't seen anyone living—at least not for very long—since this whole thing kicked off. I..." He paused. His Adam's apple bobbed as he considered it. "It's a long story. Where's everyone else? Did they—?"

"They didn't make it."

"I am so sorry to hear that," Demont said, flopping down in a swivel chair. "We lost a lot of good people here today. I found Lola, the woman you met at Guest Services..." He shook his head. "She was one of the happiest people you could ever meet. She'd just rescued three English bulldog puppies, always wanted to show me their pictures. I found her in the staff room, her wrists slit. She'd bled all over my lunch."

"Sorry for your loss," Lilian said.

"Thank you. We've all lost people today. The tragedy... it's unquantifiable."

The chair squeaked and wobbled as Ben sat, sighing heavily, glad to be off his feet. "So now what? We just sit around and wait for someone to rescue us?"

Lilian glared at him as she sat in the chair between him and Demont.

"I think I saw cards in one of these drawers," Demont said.

"Cards? We're gonna play Go Fish?"

Demont shrugged. "Something to kill time."

"Don't mind him," Lilian said. "He's got a bug up his ass about this place."

"You think I like being stuck inside a prison?" Demont asked. "Oh, I found something you might be interested in." He crossed the room to where his glasses lay beside the pile of shiny silver fabric. He didn't pick it up, merely directed their attention to it.

Lilian stood and approached him. "What is that?"

"We call them keeper suits. I found it in a locker at Guest Services. I was thinking it might come in handy for one of us if that barrier doesn't hold for some—"

He backed away as Lilian picked it up and let it drape. It looked like one of the silver suits Ben had seen at the gallows exhibit, the kind that crackled with static electricity when the ghosts had touched it.

Lilian held the suit up, looking it over with obvious distaste. "How do we decide who gets it?"

"You know, I would feel a lot better if one of you wore it. Seeing as it's my job to protect you from this place."

"Not anymore," Ben said.

Demont chuckled bitterly. "Yeah, I don't suppose I'll be getting paid this week, huh?"

Rather than putting it to a vote or drawing straws, Ben suggested, "I think Lilian should wear it."

She looked up from the suit with surprise. "Why me?"

"Because you're—"

He saw her fists clench on the material. "If you say because I'm a girl, I swear to God, Ben..."

"I wasn't gonna say that," he replied, vaguely annoyed that she would even think it. "You're smarter than me. You've always been better with directions, with planning. If anyone's gonna get us out of here, it's you."

"But you're better with close-range weapons," she said.

Demont gave them a quizzical look. "Wait, how do you know that?"

"*Infinite Zombie*," they both said.

Lilian gave the shiny fabric a dubious look, obviously thinking it over. "If I wear it, you have to promise you'll get behind me when I tell you to, Ben."

Ben nodded. "You're the boss." He had no intention of following the order. If it came down to one or the other of them, he'd already decided to put her life above his. He'd already died once. And the possibility of him dying inside Garrote House had always been likely. He'd known today might be his last day alive for weeks now. He'd come to terms with it. To destroy that house, to protect anyone else from suffering the way he had or worse, he'd always meant to put his life on the line.

And if they made it out of this prison alive, he might still get his chance.

"Promise me," Lilian said sharply.

"Okay," he said. "I promise."

"As far as I can tell, the park is running on the standby generator," Demont said as Lilian slipped her feet into the legs of the keeper suit. "There's still hot and cold water running in the fountains and washrooms and the exits are all secure. We'll know for sure if those big lights on the wall don't come on when the sun goes down, but hopefully we won't be here that long."

Ben agreed with a vigorous nod.

"Backup power means the Recurrence Field is still functioning, thank God. But that's a double-edged sword. On the one hand, it's keeping the spirits contained within the park. I saw a couple of them make a sprint for the outer wall—it didn't end well for them. Blasted right into oblivion. On the other hand, it's still amplifying the dead energy *inside* the park. That's why the bad guys were able to kill so many people in such a short span of time.

"Another problem: the backup generator powers the main park program in the control room, but it looks like most or all of the ESPs are offline. That's why the ghosts are free to move around and how the virus keeps spreading. Amplified dead energy, unbound from restrictions, means bad news for the rest of us." He stressed this point with a dark smile before moving on. "Now the only way to prevent Garrote from chomping up every ghost in the park like Mr. and Ms. Pac-Man is to shut off the backup generator. But it's in the electrical building on the other side of a thirty-foot stone wall, and even if we could get to it, shutting it down might release all the ghosts into the outside world."

"Damned if we do, damned if we don't," Ben said.

"That's a good way to put it."

"So, we wait," Lilian said, zipping up the suit. It was bulky and looked a little awkward. She twisted back and forth. "How do I look?"

"Good," Ben said a little too eagerly.

"It doesn't make me look fat?"

"Jeez, Lilian. Get a grip."

"I'm just asking."

Demont eyed them a moment. "What I'm trying to say is, the longer we wait here the more likely we are to attract Garrote's attention. He's searching for survivors. The more ghosts he can infect, the more eyes he has on the ground. You have to understand, *this*—" He waved a hand around expansively. "—what I've made here was always meant to be a temporary solution. The fact of the matter is, unless somebody outside happens to shut off the generator or tear down the damn wall, ain't nobody getting in here to save us. I hate to say it but we are on our own."

"What if we go back to the control room?"

Demont shook his head. "That's the first thing I tried. The whole building is locked down, not to mention heavily guarded. Garrote's got ghosts patrolling the entire area. There's no way he's gonna let us walk in there and shut everything off. That would mess up his plan."

"Then we're screwed," Ben said.

"Not necessarily."

"Why not?"

"It's still possible we could make a mad dash for Garrote House and get out before anyone even notices. See, this park lies on a series of underground maintenance tunnels—"

"I thought you were gonna say a Native American burial ground," Ben said.

Demont fixed him with a vaguely annoyed glare. "The main tunnels are wide enough to drive a truck through, so maintenance can get from one exhibit to the next quickly and easily without having to disrupt the guests."

"That's smart," Ben said.

"Yeah. The larger exhibits, like the prison, they each have their own entry portals. So, we get to the tunnel, we may have safe passage all the way to the maintenance hatch. We get out of the park, get the main power back on, trap those angry mother-suckers inside the exhibits. Maybe then we could find somebody to fix the program from the outside. Hack the system. Deactivate any malevolent spirits and wipe out the virus from the ones Garrote's program has already infected."

The room fell silent as both Lilian and Ben considered the plan. "But how?" Lilian asked finally. "Even if we could make it to the house without Garrote finding us, we don't know the code to get out."

Demont's eyes glimmered and he tented his fingers against his lips.

"Why are you smiling?" Ben asked.

"Because I know the door code," Demont said.

GHOST OF A CHANCE

LILIAN AND BEN considered Demont's plan while the man himself went downstairs to check on the generator.

"I think he's right," Ben said. "We can sit around here while the generator runs out of gas or we can be proactive. Allison, Niko and Leonard—they died to find that code. Now we have it. We *have it*, Lilian. We can get out of here. We can *escape*."

Lilian still couldn't believe Demont had changed his mind after they'd come so far and survived so much to get here, a place of relative safety. She'd put so much hope in this sanctuary and now he wanted to leave. It didn't make sense to her. "We're safe here," she said.

"Maybe. But for how long?"

That was the question. How long until someone from the outside world came looking for them? Hours? And how long until somebody decided they should call the police? Then what? The cops would try the gate and find it impenetrable. Even if they could get through with acetylene torches or battering rams, or climb over it, what good would their weapons do against Garrote and his army? Even if they sent for a helicopter, there could be hundreds of survivors hiding out here. It would take hours to get everyone out in small groups. Could they survive the whole night inside the prison if they had to? Would the gasoline last that long?

She heard Demont whistle something tuneless down below. As she looked out the window a bright red streak flickered briefly on the glass. She dismissed it as an errant reflection and shook her head, trying to focus her thoughts.

"I don't know, Ben. I just have a really bad feeling about that house. Ever since it came through town... since you died and came back... I've been scared of literally everything. And now I look back on it, all that time I was scared for nothing. I've never been this close to death before, Ben. Ever in my life. I should be paralyzed by fear. And somehow, I keep running. But I can't run much longer. I'm tired. I just wanna crawl into a corner and sleep for a year. So please, just... let's wait here twenty minutes. Just twenty more minutes. Then we can go if you still want to. I just need to take a break for a bit, okay? Please?"

Ben nodded, exhaling sharply through his nose. "I'm tired too," he said. He looked at his watch. "We can wait. Let's say half an hour. If help doesn't show up before then, promise you won't fight me on this."

She made to reply but another blur streaked across the glass, catching her eye. This time it was green. "Oh my God," she said. The chair spun as she stood abruptly. "It's orbs!"

"Huh?"

"Orbs, Ben. They're inside the prison!"

He looked out the window. "I don't see anything."

"I just saw two of them float past the win—" She stopped, realizing what she'd implied. If orbs had gotten inside, it couldn't be as safe as Demont had promised. It might not even be safe at all. And she'd just proved Ben right.

So much for that break, she thought.

He stood and peered out the window. "Lil, if there are orbs in here—"

"I know, I know!"

Scowling, Ben moved to the closest security camera and pressed the power button. The screen remained dark. It didn't even make that high-pitched whine old TVs did when you flicked them on. "It's not working."

"Maybe it's not plugged in."

He moved to the next one and pressed the button. "There's no power up here. Didn't Demont say he was watching us on the security cameras?"

Lilian frowned. "I don't remember. Why does that matter?"

"Because if he—"

The door opened before Ben could finish, and he closed his mouth abruptly.

"Okay, at a guess that gas should hold us for at least a few hours," Demont said. He frowned and looked back and forth between the two of them. "Did I walk in on something private?"

"Nope," Lilian said too quickly, feeling suddenly anxious.

Because she thought she understood what Ben had been trying to say. If the security cameras weren't running Demont couldn't have been watching them. So how had he known to warn them about the door just as Ben had reached out to touch it? How had he guided them every step of the way? It wasn't possible to see through the entryway from up here without the monitors working. Beyond the inner door all she'd been able to see was about three feet of green linoleum.

Demont narrowed his eyes. "So... what have you guys decided? Do we stay or do we go?"

She had to think fast. If Demont knew about the orbs, then he knew the prison wasn't as safe as he'd made it out to be. Which meant he'd lured them with false promises. But why? His intent had never been to stay here. Almost as soon as they'd gotten here, he'd started in with talk about the hatch under Garrote House. Whatever the reason, she couldn't help but feel like they had been led into a trap.

A cold trickle of sweat dripped from her underarm within the keeper suit. As she started to unzip it, feeling suddenly smothered and much too hot, she remembered how Demont had shied away from her when she'd picked it up, looking at it the way an arachnophobe viewed a spider.

Is it possible? she wondered. She couldn't risk removing her glasses, not without alerting him to her hunch. She'd just have to figure out how to get close to Demont without him noticing. She wouldn't feel safe until she was sure she could trust him.

Ben had asked Demont about the monitors.

"That's odd," Demont said, going to his side. "They were working a few minutes ago."

"You had power up here?"

"Must have been on the backup generator. Maybe it's on a timer or something, routing power through the exhibits."

Ben frowned, seemingly skeptical. Demont's back was turned and Lilian cautiously began her approach. He glanced back and she stopped in her tracks. He saw her and took a single step back, without expression.

It's not gonna work, she thought. She had to distract him. But how?

"I saw an orb," she said, thinking fast. "Two of them, actually."

"They must've gotten in somehow," Ben said, standing on his toes to look out the window again.

"That's not possible."

Ben pointed. "Look, there's one!"

"Ben." Demont glowered at him without even bothering to glance out the window, and sat back down in the chair. "They couldn't have gotten through the barrier. It's one-hundred percent ghost-proof."

"What about the ceiling?" Lilian asked. "It's glass."

"Right, the glass ceiling," he said, rolling his eyes. "Even if you did see an orb, they're harmless."

"We saw one possess a chocolate bar wrapper," Ben said. "It floated for like three feet."

"Ooh, the Case of the Haunted Snickers. That'll sure stump Scooby and the Gang." Demont looked at him over his tented fingers resting against his nose. "Ben, I don't know a ton about the science behind this stuff, but in my experience, orbs don't act like poltergeists. They're the butterflies of the ghost world. All they do is float around and look pretty."

Lilian said, "I don't know if they're harmless but there's at least two of them inside here and we've seen those things swarm. Haven't we, Ben?"

"*Swarm?*" Demont chuckled derisively. "Lilian, I would have *seen* something—"

She jabbed a finger toward the glass. "Just look out the goddamn window!"

Ben flinched, her voice surprisingly loud in the small domed room. Sighing, Demont rose from the chair and crossed to the window.

Ben gave her a curious look as she crept up behind Demont while Demont peered out for half a second. "I don't see any—"

He was already beginning to turn when she touched his arm with the silver glove.

The reaction was instantaneous. Her fingertips began to crackle with electricity as the flesh of Demont's elbow peeled back like singed paper. His eyes went wide and he let out an agonized howl. She half-expected him to shove her away, which would make his pain much, much worse. But he was smarter than that. He kicked the chair closest to him directly into her knees. She felt the scabs split open beneath the fabric of her jeans and staggered back in pain.

But the damage was done, Demont's secret revealed. He was dead. All this time, they'd been interacting with his ghost. Which meant they had never been safe here from the beginning.

"Stay back," Ben said, aiming the stun gun as Demont scrambled back onto the counter and shrank against the glass.

"Wait!" He held up his hands. "I can explain!"

"You'd better!" Lilian said, holding out her hands within the shiny silver suit like a wizard.

"Just please," Demont gasped, "don't hit me with that... thing." He nodded toward the stun gun. "Okay?"

"Then don't move," Ben countered.

The ghost groaned, grasping his elbow. Already the skin was recreating itself, knitting together until his elbow appeared solid again. "Guys, it was just an accident. I swear to God. I was setting up the perimeter, I'd just gotten the last wire in place and turned on the juice and *bam*, fifteen-thousand volts shot right through me. One second, I was on the ground in the worst pain of my life and the next, I was floating above myself. Like an out of body experience. Only when I tried going back my body wasn't having me." He shrugged, as if the idea that he'd just died meant nothing. "So I just dragged it out of sight," he said. "By then, you guys were already on your way here. It wasn't like I could call you back and tell you not to bother."

"You need us to help you escape," Ben said, giving Lilian a fearful look. "Don't you? You can't open the hatch yourself. The electricity would fry you, like the keeper suit just did."

The ghost nodded solemnly. "Because Garrote is coming. We can feel it. As soon as that generator runs out of gas, him and his minions will get inside and I won't be Demont Hudson anymore. I'll just be part of his army. His hivemind. Just another ghost for his collection."

"But you're safe here, why would you want to—?"

"Because there's more of us, Ben," Demont interrupted. "We've been hiding here since it all went down." He directed their attention toward the window. "Just look."

Demont shied away again as Lilian sidled past him to the window. He put two fingers between his lips and whistled, and within moments the area below filled with ghosts, materializing out of thin air. There had to be dozens of them, all looking up at the control tower windows.

"Holy shit!" Ben said.

Lilian thought it was an understatement.

"DON'T WORRY," Demont told them. "They won't hurt you."

"How did they get in here?" Ben asked.

"They were here the whole time, hiding from *him*. I just hadn't seen them. I was too busy running around, trying to play the hero. But my plan worked. You can see that. Nothing is getting in here so long as that generator keeps running. We just can't get out."

Ben looked down at the ghosts crowding the Circle of Death. All eyes were focused on the three of them in the tower, although he wondered if they were able to see through the tinted glass. Still, his heart thumped heavily, fear pumping adrenaline through his veins. His medication had worn off. If things got worse—if the ghosts rushed the tower or the generator ran out of gas and Garrote and his army broke in—he'd need to take another pill. He'd never taken more than the recommended dose before and wasn't sure what the side effects might be.

Breathe, Ben, he told himself. *Just breathe. In and out. Don't hyperventilate.*

"So why do you need us?" Lilian asked.

"Like Ben said, as spirits we can't interact with anything electrical without experiencing... technical difficulties, let's say. My guess is that it's part of the coding, so the electrostatic precipitators will function as barriers. It certainly doesn't work like that out in the real world. The only barriers out there are of our own making. Psychological barriers, binding souls to objects and places until we're able to move beyond those barriers, beyond the physical plane."

"But you're not like the rest of them." Lilian nodded toward the window. "You weren't part of the Ghostland program. So why should the electrostatic effect you?"

The ghost shrugged. "The program is still running, still detecting dead energy. That's how it works, I guess. The second my soul—spirit, dead energy, whatever you want to call it—the moment I left my body a new algorithm must have been added to the code. That's all I can figure. I can do just about anything I could before and then some. Like move things without touching them. And I can walk through walls." He smiled briefly. "Being dead does have its advantages."

Ben wasn't convinced. Something didn't quite gel with Demont's story and what Sara Jane had said about the Ghostland program. He just couldn't put his finger on it. "If we help you escape," he said, "what's to stop your friends from killing us the second we get you out of here?"

Demont let out a sharp laugh. "Them? They're harmless. We just want to leave this place and never come back. Move on to whatever comes next. You know I never fully understood the implications of what that evil woman created here until I met these people. Ghostland is a modern-day plantation. Think about it. These poor souls are all prisoners here. They perform tasks for zero compensation. They're tortured—"

Lilian frowned. "How do you mean, tortured?"

Demont gave her a hard look and spat out his reply. "That 'Recurrence Field' forced them to relive their deaths over and over, for *entertainment*. If that's not torture, I don't know what is. Everything that woman did to us—to them—that's the *definition* of slavery."

Us, Ben thought. Demont was already acting as if he was one of them. He'd sure changed his opinion about Ghostland since their ride from Guest Services to the control room. Maybe death had made his thoughts erratic. Ben could certainly sympathize. Death—even the near-death he'd experienced himself—changed everything. It could awaken things in people they never could have comprehended or even dreamed of before.

Lilian seemed to have arrived at the same decision, giving him a slightly troubled nod.

"Okay," he said, turning to Demont. "Let's get out of here."

LILIAN HAD MADE Demont promise he would keep the other ghosts a safe distance from them before they set foot outside of the guard tower, but she still felt a tremor of fear as they headed down to the Circle of Death. The eyes looking up at them seemed so haunted. She saw sorrow, pain, confusion. But unless she was reading it wrong, she also saw a glimmer of hope.

Lilian counted heads. There were twenty-three in all, including Demont. She spotted a circus clown with much of his flesh burned off, alongside two Victorian street urchins. Behind them stood ghosts she hadn't seen before, or at least didn't remember. There was a pirate whose skin had gone grey and bloated, draped in lengths of seaweed that moved like snakes over his dripping garments, and a Civil War soldier who kept absently slopping the bloody loops of his intestines back into his abdomen. Beside them was a Chinese man wearing a newsie cap and suspenders to hold up his loose denim pants. His body had been flattened at the shins and chest as if by a vehicle, stretching him taller than the rest by at least a foot. To their left, a football player held his head in a helmet under his right arm. To their right, a ballerina stood *en pointe*. Her heart had been carved roughly out of her chest, leaving an open cavity, and blood stained the ruffles of her pink tutu. At her feet a golden retriever with bare patches of mange in its fur panted and foamed at the mouth.

There were others, many others, and Demont smiled and spread his arms wide to this diverse group as he reached the floor. "My people!" he said. "We are getting out of here!"

The ghosts broke into applause, cheers and whistles. The dog wagged his tail happily. Despite her fear, Lilian smiled. These restless spirits would finally find freedom. It was an exhilarating moment to be part of, ghosts or not.

"Step back, please, step back," Demont said, parting the crowd. They were already backing away from Lilian as Demont led them through, eyeing her with a mix of fear and fascination, obviously aware of what the keeper suit could do if they got too close. She supposed some of them might have suffered at the hands of a keeper before the fall of Ghostland, and she was struck with a twinge of guilt even though it hadn't been her decision to wear it, and it had only been for her protection. At her side, Ben was acting like a reluctant celebrity, greeting and nodding at ghosts as he passed, not afraid in the least.

"One of you will have to shut off the generator," Demont said as they approached it. "Much as the idea of that troubles me. Once it's off, the two of you run like the wind down that wing there for the back exit." He pointed to a barred archway. There were signs on either side of the tunnel but Lilian couldn't read them from the distance. "We'll protect you as best we can if Garrote and the others get in before we get out," he said.

Nods among the crowd greeted this.

Ben raised a hand. "I'll turn it off."

"We have ourselves a volunteer." Demont ushered him toward the rumbling machine. Lilian followed until Ben stopped suddenly and turned to face the ghost.

"Wait a minute, you didn't get shocked by the generator, did you?"

"What?" Demont squinted, then shook his head. "No. It was the bars. It was an accident, like I said."

The dog barked behind Lilian. She almost jumped out of her skin.

"Be quiet, Freddie," Demont said.

"But why did you touch the bars after—"

"*Ben.* Do you wanna argue all day or are we gonna get out of this goddamn place?"

Ben shrugged. He approached the generator.

"You see that key?" Demont asked. "All you have to do is turn it counterclockwise."

"That's it?"

"That's it."

Ben reached for it. "Here goes nothing," he said. He looked back over his shoulder at Lilian, who nodded for him to proceed.

Then he turned the key.

PRISON BREAK

THE ASSAULT BEGAN as soon as the generator stopped rumbling and the hum of electricity faded to nothing. The glass dome shattered and dozens of orbs led an army of soul-suckers down through the opening. The ghosts below scattered in the rain of glass, some vanishing, others floating away. Unable to do either, Ben dashed for the barred archway Demont had pointed out. It wasn't until he got halfway there that he could read the signs on either side: EAST BLOCK CONDEMNED ROW II.

Under normal circumstances it would be a place to run away from, not toward, and seeing this was where Demont had directed them didn't exactly inspire confidence. But with their sanctuary compromised, there was little choice. He ran ahead, glancing back to see that Lilian was following him. The dog dashed past him and headed right through the bars, barking down the darkened tunnel until he vanished. The ballerina leaped and leaped from toe to toe, her form so perfect Ben was surprised she didn't twirl.

Lilian cried, "Watch out!"

Something shot toward him out of a darkened cell. He ducked, and the rusted metal panel whipped over his head. His backpack jerked roughly into the air, tugging on his shoulders hard enough to pull his feet off the ground.

He hurled the backpack off his shoulder as it began to unzip itself, the flap falling open like a tongue and its orb-possessed contents floating out: a crumpled granola bar wrapper, carefully folded tissues that bloomed and shaped themselves into origami, the water bottle, his pills, his Rex Garrote paperback, the pages flapping like an angry bird—

The lighter fluid!

A soul-sucker landed on the dirty Victorian boy to the right of him. The boy struggled as the creature's black tendrils looped around his slender wrists, screaming soundlessly as it sucked his soot-streaked face into its darkness, like matter drawn into a black hole.

Ben wanted to help but he knew he'd lose his backpack if he let go of it now. He thrust a hand in up to his elbow and unzipped the secret pocket, while the cover flapped against his arm like a chewing mouth and its contents battered his arms and torso. He managed to snatch the can and matches and tucked them swiftly into his pocket, hoping they hadn't already been possessed—but as he did, he saw his pills had floated out of his reach.

The boy was beyond saving. The soul-sucker had completely enveloped him and was undulating where he'd stood as if digesting a large meal.

Ben swung out with the bag, holding it by a single strap, and barely managed to knock the pills out of the air. They hit the ground and the plastic lid broke off, pills scattering across the painted concrete floor.

"Come on!" Lilian shouted back at him from the doorway.

"I need my pills!"

He let the backpack go and dropped to his knees, collecting the pills one by one with shaking fingers.

"Ben!"

"I'm coming, I'm coming!" He grabbed two more and stuffed them into his pockets, then caught up to her. Both of them flinched at a loud metallic *clang* that echoed throughout the cells. They looked up as another clang resounded. And another. Another. Lilian's eyes went wide as all around them the cell doors swung open and slammed against the bars, one clang following the next around and around the Circle of Death, and the monstrous prisoners they housed—tattooed hulks and wiry, gray-fleshed lifers—stepped into the light of day.

Ben and Lilian didn't waste another second. They ran toward death row, and the headless football player and flattened railroad worker opened the bars as they approached. They dashed through the archway and into the cool, dank tunnel. The footballer pulled the gate shut and Ben kept running. Lilian had sprinted ahead.

Lightbulbs in metal cages hung from the ceiling but without power the tunnel grew darker the further he ran. The dog's excited barks echoed from somewhere in the dark up ahead. He saw Lilian pause near the far end of the tunnel with her hands on her knees to catch her breath.

Ben spared a glance back and stopped. The dead prisoners stood within the Circle of Death, several of them fighting off the Swarm, the orbs making the objects from his backpack swirl around them, with various other detritus from within the exhibit.

Lilian screamed. Hushed voices calmed her. Ben ran for her, as fast as his heart would allow. He couldn't see any of them up there in the dark. They must have already rounded the corner.

A moment later he found himself in a small lobby, DEATH ROW INTAKE painted on the wall. Cubby holes in a small inner room to his left were stuffed with prison jumpers, towels, shoes and toiletries. A dead body lay propped up against them, and the shock of sudden recognition struck Ben. This was why Lilian had screamed. It was Demont lying there, his face gray and lifeless. And even though Demont had already warned them, seeing his body there was a bitter pill Ben had swallowed many times today: any one of them could die here, at any time.

No one was safe.

The heavy metal door to death row lay open just ahead, not ten feet away. But something about Demont's corpse troubled him. With a glance back the way he'd come to make sure the prisoners had yet to follow—they stood in the archway, watching, shifting silently, the soul-suckers apparently defeated —Ben forced himself to approach the body on the floor.

Demont's features had contorted with fear and his eyes looked horribly bloodshot. Blood had oozed over his left ear and down his neck from a gaping wound on the top of his head. But Demont hadn't said anything about a head injury, and there were no burn marks on his hands, no prominent veins or blackened fingernails to indicate electrical shock. Ben had seen too many autopsy photos to count from the morbid internet sites their former friend Brody had been obsessed with. What he saw here was not an electrocution victim but a man whose skull had been caved in, likely with some kind of blunt object.

Why would Demont lie? he wondered.

There was no time to guess. The orbs had started into the tunnel, casting a sinister Christmas tree glow on the ceiling and walls. Whatever really happened to Demont, Lilian was in death row with him. Whether they still chose to help him and the other ghosts to escape was something to worry about later. Right now, getting out of this hellhole before the rest of Garrote's army found them was more important.

He dashed through the doorway into death row. Cells with a single slit at eye height and hinged slots for meals lined the walls on either side. At the end of the cell block Demont stood just inside the door, ushering him forward. Lilian stood beside him, desperation in her eyes, holding the door open. "Come on!" she said, waving her free hand frantically.

With no other choice, Ben hurried ahead. He just had to hope Demont had lied for a good reason. He slipped through the door she held into a large room with several closed doors, each labeled according to its purpose: CHAPLAIN, WITNESS ROOM, DEATH CHAMBER. Directly ahead was a door with a red EXIT sign above it. Lilian pulled the door shut behind them. It clanged heavily, but Ben knew it wouldn't afford them any protection from the ghosts at their heels. They had to get out of here, into the open. This place was a death trap. It could come at them from all sides, even through the walls.

"Streamroller! Mister Lim!" Demont shouted to his people. The football player and the railroad worker stepped out of the group, ready for orders. "Guard the door until we've gotten the children through!"

Demont gave them both a pat on the shoulder and a brusque nod. Mr. Lim[1] cupped his left fist in his right palm and bowed slightly. Steamroller[2] merely grinned, his severed head still tucked under his arm. Both men took up a stance on either side of the door, waiting to defend their temporary stronghold from the monsters at the door.

The group parted again as Demont led Ben and Lilian to the exit. "When I open this door, you two run like hell to the end of the hall. We'll make sure it's open when you get there."

Ben and Lilian looked at each other and nodded.

"Enemy at the gate!" the Civil War soldier shouted, stepping halfway through the inner door and slopping his guts back through his long blue jacket. Without warning his body began to shudder, his facial features twisting until it was Garrote standing there, halfway through the doorway, his innards hanging out of the Confederate uniform. He grinned and began to sing: "'Well, I wish was in the land of cotton—'"

Growling, Steamroller charged the door. As his neck stump and shoulders slammed into Garrote his body began to jitter and he lost hold of his severed head. Even as the helmet hit the floor and rolled toward Ben, he could see Streamroller's face changing shape and complexion, growing a brushy, dark mustache and a malicious grin.

"Ooh! Fumble," Garrote's severed head cried from within the helmet.

"They're infected!" Demont shouted. "Everyone out!"

But it was already too late. Ben stood there watching as ghost after ghost morphed into another Garrote clone. He couldn't help feeling like everything they'd done, every moment, every failure, every struggle, all of it was for nothing.

Lilian grabbed his hand and jerked him roughly forward.

"Go!" Demont shouted. He'd drawn the heavy door open and was waving at them urgently. As they headed for him, uninfected ghosts floated by on either side. The dog dashed by at Ben's feet, panting. As the ballerina leaped by her body began to shudder. In mid-jump she twirled, her face shifting, widening, flattening out. She landed sprawled out on the floor and as Ben hurried past it was Garrote's hand that reached out to grab him, the writer dressed in the blood-drenched tutu, his hairy body stretching its pink fabric.

"Wanna dance?" Garrote growled, and laughed maniacally.

Ben leaped over the grabbing hand. He tripped over his own foot, took a few sprawling steps forward and likely would have fallen flat on his face if Lilian hadn't caught him by the arm and pulled him onward. At the end of the corridor, Demont reappeared and hauled open another heavy door. Sunlight flooded the room. The dog parked itself at Demont's feet and barked, foam flicking from its lips.

"Quickly!" Demont shouted, pure terror in his eyes as more and more of his friends succumbed to the virus.

Lilian rushed out the door. Ben threw all of his remaining energy into spurting forward. He dashed out into the waning late-afternoon heat, then lurched forward with his hands on his knees and retched. He spat between his feet and caught his breath, barely managing not to puke.

"Ben," Lilian said. She was squinting down at him, her hands on her hips. He noticed she'd barely broken a sweat. "Can you make it?"

"I'll be..." He took a deep breath and exhaled sharply. "*...fine.*"

Freddie the dog bounded over to them. Demont didn't even bother to pull the door closed. He vanished from the doorway and reappeared at their side.

"We have to go," he said. "Now."

They stood under the shadow of something large with angles as sharp as teeth. Ben finally looked up at the house. It stood on a small rise across a short promenade, dark and imposing, silhouetted against the falling sun. A wall made of moss-covered stone surrounded the grounds. Ivy had crawled up the stone pillars and weaved through the rusty metal arch and the letters spelling out GARROTE HOUSE in wrought iron.

Rex Garrote himself stood below the rusted arch, grinning. He began to clap slowly as he moved toward them. "I must say, I am quite impressed," he said. "You've come a long way, babies. But now the time has come to say goodbye."

"You can't stop us, Garrote!" Demont shouted, his hands clenched into fists.

"Oh no? Look around you. These are my people now."

As he spoke dozens of ghosts shimmered into view, surrounding them in a wide semicircle, with several soul-sucking shadow creatures weaving through the crowd. To Ben it looked like a costume party in Hell. He took Lilian's hand. She looked fearless but she was shaking from head to toe.

Demont turned to them. "It's not too late to save these people," he said.

Garrote chuckled. "And just how, pray tell, do you plan to do that? I'll let you children in on a little secret. There is no service hatch in my house. I created that myth to lure you here—a very cunning trap, if I do say so myself. I'm afraid the only way out for the two of you is in pine boxes."

"He's lying," Demont growled.

Garrote's head twitched to the side. "Perhaps. Children, did you know you're in the presence of a true criminal mastermind? Isn't that right, Mr. Hudson—or should I say, Mr. Betruger?"

Lilian frowned. "What's he talking about?"

Demont shot them a fearful look. "It's not true."

"Truth is relative," Garrote said. "You of all people should know that. You see, in life, Lance Betruger began his career as a banker and quickly graduated to con artist, bilking thousands out of their retirement plans. Stealing their dreams. Of course, it was only a matter of time before the authorities caught up with him. Far too cowardly to take his own life, Betruger wound up in a twenty-year stint at Fontaine County Correctional. Which is where your friend Mr. Hudson ran into him, much to his detriment.

Or rather, where Betruger ran Mr. Hudson headfirst into the bars of his cell, crushing his skull."

"He lies!"

Garrote smiled, continuing unabated. "Betruger, you see, is what we paranormal enthusiasts call a 'doppelganger,' and like all doppelgangers he's able to change his form to whatever tickles his fancy."

"You shut the hell up, Garrote!"

"Although he doesn't like to be reminded of his current status—*sans corps*, as the French say. In point of fact it makes him quite upset, doesn't it, Betruger?"

"*You shut your filthy mouth!*"

Lilian shot forward, crying out in rage. She grabbed Demont by his shoulders and squeezed. The ghost—Betruger, Demont, whoever he really was—howled in agony as his clothes and flesh singed away from the electrified gloves. He tried to twist free, tried to turn and face his attacker, but Lilian held him firmly.

Ben felt the sting of betrayal just as sharply and stepped forward with the stun gun. He depressed the trigger. With a single jolt, the man wearing Demont's face fizzled into oblivion.

The ghosts shifted silently, reverently. In the silence, Garrote began to chuckle.

"Bravo, children!" he said. "It would appear I've underestimated the two of you. Here I'd thought you'd survived upon the kindness of strangers, but you've really shown true grit just now. *Chutzpah*, you might say."

"Fuck you!" Lilian said, spit flecking the inside of the clear plastic mask.

"My, my, can't pay a girl a compliment anymo—"

"Shōki!" Ben shouted.

For an instant Garrote's eyes burned with fury. Then he smirked, like a man who knew a secret. "What do you know about that, *shy boy*?"

The name sent a shiver of fear up Ben's spine but he wouldn't allow himself to back down. He'd gotten to Garrote. He needed to keep needling him, make him flustered, distracted. "I know what you're doing," Ben said. "I've read all of your books. You're collecting ghosts, like the Chung Kwei. The Shōki."

Garrote's lips rose in a half-smile. "My number one fan, eh? How touching."

Lilian leaned close and whispered, "What are you doing?"

He ignored her. "Uh-huh," he said. "I've read all of it. All your interviews. Your non-fiction. Rex Garrote, the most terrifying man in the world."

"Tell me, Benjamin Laramie... do you sincerely believe flattery will have me spare your life? You said it yourself, you've read my *Shōki*. You know as well as I do, only one man lives to see the end."

"Except you're *not* alive, are you? You're not even a ghost." Garrote flinched at this, and Ben kept needling, taking a cautious step toward the dead man. "Without your army, you're nothing. *Less* than nothing." He remembered what Sara Jane had said, and hurled it like a weapon: "You're a computer algorithm. Ones and zeroes in the shape of a man."

"A pretty ugly one too," Lilian added. "And old."

Garrote stared at them a moment, and Ben realized for the first time that for all the realism the programmers and 3D modelers had put into the writer's avatar, he'd never blinked. Not once. It was enough to make Ben believe maybe there was a chance to survive this after all. They truly were fighting against an algorithm, and without the internet, computer programs remained subject to the boundaries of their hardware.

Finally, Garrote laughed. "Really? Is that what you think? My appearance aside, do you truly believe I pose no harm without my people?" He tutted. "In here, I am a god. I control every aspect of this park," he said calmly. "I could kill you with an errant electrical wire. I could twist your headsets like vices to crush your skulls and paint my house with your brains. I could choke—"

"Blah blah *blah*!" Lilian said. "We know about the red water, Garrote! We know about the dark red water."

Ben turned to her, wondering how she'd known about the red water if she'd never read *Shōki*. The oceans had run red with the blood of millions and the sky had filled with ash.

No matter how she'd known about it the effect was instantaneous. Garrote snarled, showing his too-white teeth, and raised a hand toward them in anger —

Before he could attack, a pounding roar filled the air. He howled at the sky in frustration, like a man cursing God, as a helicopter rose above the house, thundering loud enough to shake the earth. The glass bubble-dome glimmered in the sun as it turned toward them, hovering thirty feet above the ground.

Rescued, Ben thought. *Finally.*

He waved his arms desperately. Lilian joined him, shouting for help.

A police officer leaned out the open door, holding a megaphone. "THIS IS THE WASHINGTON COUNTY SHERIFF'S DEPARTMENT," the man's voice boomed. "PLEASE SEEK SHELTER IMMEDIATELY. WE WILL BE EVACUATING CIVILIANS IN ORDER OF SEVERITY."

Lilian lowered her arms. "What? No! We need help!"

"Help us!" Ben cried, still waving, refusing to give up.

"They can't see the ghosts," Lilian said, realizing just as he did that the police weren't wearing headsets. "They can't see we're in trouble!"

"*TAKE IT DOWN!*" Garrote roared.

In an instant, the ghosts launched into the air to attack the police. A mental patient reappeared alongside the cop hanging halfway out the door, grabbed him by the collar and tore him from the helicopter. He fell screaming. Ben winced at the sound of his bones breaking as his body twisted sharply over the wall, audible even over deafening sound of the rotors. The megaphone smashed to bits below his broken corpse.

The Swarm struck the helicopter next, battering it on all sides. The aircraft shook and dipped, the pilot struggling to maintain control. A one-eyed pirate shattered the dome with his hook hand and grabbed the pilot out of her seat. He lifted her, the hook caught under her chin, until her head caught in the whirring blades and tore off. Ben and Lilian covered their heads from a rain of shattered glass and a mist of the pilot's blood. Her body followed a moment later, sprawling on the concrete at Garrote's feet.

Garrote was watching his concerto of chaos with malevolent delight. Sparks flew from the cockpit as his minions tore out its instruments. The metal warped as the shadow creatures battered its hull and bent its tail.

Lilian grabbed Ben's shoulder. "We have to go now," she said.

And then the helicopter fell.

She ran for the house. Ben followed.

The aircraft hit the ground and exploded with a wave of heat that thrust them forward. Ben ducked, expecting a volley of shrapnel, but he didn't dare stop, just kept running headlong for the gate as burning chunks of twisted metal sailed over their heads. Garrote was so absorbed in the mayhem he'd unleashed he didn't notice as they dashed past him. And even though he knew it wasn't possible, Ben was sure he saw flames dancing in the writer's eyes.

They passed under the wrought iron arch and Ben looked up, finally getting a good look at the house that had plagued his nightmares for the past four years. The same ivy-covered walls, the same bricks, the same gables. A classic haunted house in every sense of the term. He hoped, with luck, it would be the last haunted house he would ever see.

Garrote roared behind them, realizing his mistake. They turned to see him charge the opened gate. As he passed under the archway his body was jolted by an invisible force and thrown backward several feet.

"The ESPs!" Lilian cheered, pointing at the poles standing against the stone pillars. "They can't get in!"

For now, Ben thought.

He was certain it wouldn't be long before Garrote would find a way inside. Garrote had said he controlled the park and Ben had no reason to doubt him. It would simply be a matter of finding the right algorithm and deleting it from the program. After that, the Shōki and his army would storm the house and that would be the end of everything.

PART 4
THE LAST HAUNTED HOUSE

Garrote House is not only haunted by the spirits of former residents, but is believed by some to be plagued by an entity many paranormal investigators and spiritual mediums have called "demonic," or "cursed." While the ghosts you will encounter inside are real, there is no concrete evidence that any such malignant supernatural force exists within its walls.

— *Know Your Ghosts:*
A Guide to Ghostland

It was impossible to tell where the house ended and the writer began; they were one being, joined at the heart and the hearth, the bones and joists. Separating one from the other would destroy them both.

— Rex Garrote, *The House Feeds*

THE HOUSE WAKES

BEN STARED UP at Garrote House[1], thinking about truth in fiction.

Having read *The House Feeds* and its two sequels countless times, he knew navigating the house wouldn't be quite as easy as it seemed. In the books, the house was sentient. It could change its dimensions, its shape, even the position of its rooms at a whim. He doubted the real house would be able to play the same tricks, but if Rex Garrote had been planning his attack since Ghostland had been only a vague concept with no real science behind it, he was likely to have made certain that escape would be as difficult as possible for anyone left alive after the initial assault. This was Ghostland's final exhibit, the capstone of his theme park and his long career in horror. It was the long-awaited dramatic conclusion to Rex Garrote's *House* series, left unfinished after his suicide. And with any luck, an abrupt end to his afterlife.

Lilian stopped suddenly and turned back toward the gate. "I can't see them," she said, squinting behind her glasses.

Just outside the gates the twisted and blackened hulk of the downed helicopter burned. Beyond that, the park appeared deserted.

"Good," Ben said. "Maybe they're gone."

"I doubt he'll just let us leave. We got to him back there. I think we really scared him—"

"Yeah, how did you know about the red water?"

"How do *you* know about the red water?" Lilian shot back with suspicion.

"I read the book," he said.

"What book?"

"*Shōki*. If you haven't read it how do you know about the red water?"

"It's what Allison said," Lilian said. "Right before she died, she said to promise we wouldn't go 'down into the dark.' She said, she could see him—I guess she must have meant Garrote—waiting for us in the water. 'It's so red, why is the water so red?'" Lilian recited the words almost mechanically, then uttered a strangled sob. "And then she died, just like that," she finished, and grabbed Ben in a stiff embrace.

He hugged her back, scowling over her shoulder, trying to make sense of what she'd just told him. "That's strange," he said. "Why would she quote one of his books? Her last words, and that's what she says?"

"I dunno," Lilian said, letting him go. "People say weird stuff when they die. My mom said Nana asked her if the roast was burning right before she died, and she was already in the hospice by then."

"That is weird," Ben said. But there was something more to what Allison had said than that. The thing about "waiting for them in the dark" wasn't in *Shōki*, as far as he knew. And why would she have quoted the book as her dying words? It just didn't make sense.

He made to mention this when a message flashed on the inside of his glasses. It had been so long since the last one he startled and swatted the air in front of him, thinking an orb had fluttered close to his face, before he was able to focus on the message.

HEADSET DISCONNECTED

"*No*," Lilian said, shaking her head.

The words flashed on the inside of his glasses twice more and disappeared.

"*No no no!*" Lilian cried, spinning in a tight circle, trying to look everywhere at once.

Ben put a hand on her shoulder. She jerked away from him.

"We have to keep going," he said.

"But we can't see them!" She looked out again through the gate. "*I can't see them!*"

Ben couldn't see them either. But it was impossible to tell if Garrote and his minions had left or were simply no longer visible with their glasses not functioning. Whatever lies the writer had told them, he hadn't lied about one thing: within Ghostland's walls he was in control. They wouldn't be safe until they got through that hatch, and without the security code the chances of doing so were slim.

"There's nothing we can do about that now," Ben said. "We have to get inside, get to the service hatch—"

"*There is no service hatch!*" she screamed, her eyes wild with fear.

"We don't know that. Garrote was trying to get inside our heads, that's all. It's too late to turn back. We have to stick to the plan. There's no other choice. We have to keep going."

She stared at him, tears welling in her eyes. He thought she'd never looked so beautiful, so vulnerable, as she did right now, and he wondered if that said something about him, that he'd never loved her more until fear and exhaustion had broken down her defenses, stripping down her facade of teenage apathy and constant aggravation.

"You can do this, Lilian. We've made it this far, all the way here. We can't give up, not now."

Her lower lip trembled. Then she clenched her jaw and nodded. "You're right." She nodded again, more forcefully, and wiped away standing tears. "We have to stick to the plan."

They turned back to face the house. The sun had dipped below the westernmost gable, and the grounds beneath its shadow were caught in the haze of early evening. It would be full dark in another hour or so. Ben had always wanted to spend a night inside a haunted house, but he knew if they were still inside Garrote House by nightfall, neither of them would make it out alive.

Sara Jane Amblin had compared a large mass of dead energy to a nuclear meltdown. There had been so many senseless deaths at Ghostland in such a short span of time, Ben suspected the fallout from today's meltdown would linger for centuries, long after they bulldozed the park and its exhibits and salted the earth below.

"Ready?" he asked.

Lilian nodded, and they started up the cobblestone path, heading up the small rise lined with gnarled, dead trees, the branches creating a canopy. As they reached the porch steps the front door creaked open.

"Jump Scare City," Lilian said nervously, an oft-used reference from when they were younger, watching horror movies together in the dark and pretending they weren't afraid. Neither of them bothered to pretend now. They were scared out of their minds, and didn't care that the other knew it.

Ben took the steps cautiously and peered inside. In the dying light the house looked warm and inviting, possibly illuminated by candlelight. He knew it was a trick, that the house was far more dangerous than any of the other exhibits thus far. No one had seen Garrote's ghost, but there were many other tortured souls attached to this house. Their dark histories had been woven into Rex Garrote's *House* series. Some readers had suggested all of his later novels had been based on reality, that the house and its many ghosts had played out their stories in Garrote's dreams, or whispered them to him as he wrote on his old Remington typewriter, and he'd merely transcribed them.

Without a word, Lilian stepped through the doorway. Glancing back at the empty grounds, Ben followed her in.

However dangerous the house was, it was undeniably beautiful. Ben's mother had said it was Victorian, but the interior seemed to be a mix of Gothic Revival and medieval, like the inside of an old castle. A massive brass chandelier tinkled musically above their heads as if to welcome them into the long, high-ceilinged foyer. Stairs with ornate dark wood railings and oil lamps fitted atop each bannister rose in a moderate incline along the west wall from the front door to the far end of the foyer. Under the base of the stairwell stood a single door with a chaise lounge beside it. There was a sign on the door, but they weren't close enough to read it.

A massive stone fireplace stood beneath the second-floor balcony. A large tapestry hung from the railing above, depicting violent images from medieval battles to dragon slayings. He couldn't see the images from where he stood, Ben had seen them in photographs for an interview Garrote had done with *Playboy* in the mid-'80s. On either side of the fireplace two large abstract metal sculptures stood sentry, mad fusions of human and animal and ancient torture devices. These were the work of the previous owner, Clayton Odell[2], whose most famous sculptures had been displayed in the MOMA in New York and several other prestigious galleries worldwide, before he'd gone insane and murdered his entire family. Now his remaining art was held by private collectors, the rest of it relegated to storage houses and backrooms, safely out of the public eye.

A loud clatter caught their attention. Shadows began moving across the foyer. They turned to the door as thick metal sheeting lowered over the windows like the eyelids of a giant beast, finally allowing itself to dream.

"What's happening?" Lilian shouted over the noise.

Ben didn't answer, merely bolted for the door. He could have made it—just barely—but Lilian hadn't followed, and it would have meant leaving her behind, alone. He watched in dismay as the massive door slammed shut in his face. The lock clicked, barely audible over the clatter of the metal shutters, closing out the light inch by inch.

The big brass handle wouldn't budge. They were locked in, trapped inside the most haunted house in America. Two kids barely into adulthood, left alone with whatever ghosts remained inside. As the last two shutters clicked into place over the windows, steeping the foyer in absolute darkness, Ben once more found himself wondering if the security measures were meant to keep the ghosts out or trap the guests inside.

"Ben?"

"I'm here."

"I can't see you!"

"I can't see anything," he called back. His voice echoed in the large space. The darkness seemed endless. He could almost feel the emptiness. Were they alone in here? If not, they were completely vulnerable, completely at the mercy of Garrote House and its ghosts.

The house groaned, deep within its bowels. Ben couldn't tell if it was settling or footsteps he heard.

Finally, his eyes began to adjust. Small cracks of daylight dimly illuminated the foyer. A dark shape moved toward him from the center of the room he hoped was Lilian. It reached out and grabbed his shoulder. Warm fingers felt down his forearm until they found his hand. She grasped it eagerly.

"There you are."

He squeezed her hand.

"I thought I lost you," she said.

"I'm not going anywhere without you," he told her. Even if it had been possible to see, he wouldn't have left her alone, as scared for his own life as much as hers.

Another message flashed on their glasses, blindingly bright in the darkness. He had to squeeze his eyes closed and blink several times to focus on it.

INSTALLING UPDATE...

"Update?" Lilian said. "What's going on?"

"I don't know. But if Garrote's involved it can't be good."

When the progress bar had filled a second message appeared:

HEADSET PAIRED WITH SYSTEM

Suddenly all the lights came on around them, faux flame light bulbs in the lamps and chandelier lending the flickering orange glow to the room Ben had first seen when he'd peered in through the doors. A fire roared to life in the stone hearth.

Lilian had taken off a glove from her keeper suit and held it in her other hand along with the helmet. Her hair was sweaty, her bangs matted against her forehead. "Oh, thank God!" she gasped, letting go of his hand.

Ben wasn't sure he shared the sentiment. But he was glad not to be in the dark anymore, for however long it might last. Then he realized something. "The lights, the shutters... there must be power in here."

"Lot of good that does us," Lilian said, eyeing the big metal shutters. "You think Garrote can get inside with those things covering the windows?"

"I don't know. I hope not."

Somewhere upstairs a door creaked open. Footsteps moved rapidly in several directions, a frenetic cacophony of stomping feet.

Ben thought, *Survivors*.

But he knew they were likely alone.

And that whatever it was upstairs wasn't likely to be friendly.

LILIAN STAYED BY THE DOORS, looking around, while Ben started cautiously across the large foyer. The footsteps they'd heard once their headsets had come back online still worried her. At first, she'd thought it might be survivors, but survivors would have reacted to the lights going out and the windows battening down. Which meant there were multiple ghosts inside the house. And she had no idea where to begin searching for the security hatch.

"Where are you going?" she asked.

Ben turned at the foot of a large Persian rug below the chandelier. "We should start in the basement. If there is a hatch, it's more likely to be there. Down is always the way out in horror games."

"Whatever you say," Lilian muttered, thinking his logic was flawed and the hatch—*If it exists*, she reminded herself—was just as likely to be elsewhere. Puzzles and mazes almost never worked the way the player expected, and her aptitude for them had rarely steered her wrong in the past.

Still, she was too tired to argue. More than anything, she just wanted to go home. Sometimes the simplest solution was the correct one. She just had to hope Ben was right this time.

Lilian moved past him and headed to the door below the stairs marked EMPLOYEES ONLY. The chrome lock mechanism looked like it required a key card and the little light on its face was red, which likely meant it would be locked. She tried the handle, just in case.

"Well?"

She shook her head. "Nah. We'll have to find another way down."

"There's gotta be another way down there."

She said, "I hope so," but she wasn't so sure. The house was enormous. They could end up walking around for hours trying to find a route to the service hatch.

Unless Garrote found a way into the house first, in which case they wouldn't have to worry about finding the hatch, because they would be dead.

Ben headed across the foyer and Lilian followed. He tried the door closest to him. It opened inward with a groan and he stood there a moment looking into the room.

"This is where he died," he said, seemingly wary of entering.

She came to his side and peered into the large, octagonal library. Bookshelves lined the walls. A rolling ladder led to a walkway that accessed the upper shelves, surrounding its entire circumference. Above them, the darkening sky was visible through a muraled glass dome ceiling, the shadows of intricate stained-glass monsters—trolls, goblins and winged creatures with claws—falling over the shelves and the marble floor like a sinister baby

Ben turned with concern. "What?"

"The camera." She pointed it out.

"Shit," Ben agreed. "If Garrote can see us, we need to find another way down to that hatch quick."

"And you know Rex Garrote better than anyone," Lilian said. "I think it's time to put that useless knowledge of yours to the test. So... where would he put the exit?"

HACKED

BEN HAD HEADED straight for the collection of first edition Rex Garrote books. But after tugging on every single book there—and wishing he'd kept his backpack to take a few of the more precious volumes home with him, if he ever made it home—he had no idea what to do next.

"Nothing, huh?"

He shook his head. "I thought for sure it would be one of these." He pulled out a limited-edition hardcover of *Shōki* with the original samurai mask cover and flipped it open. It was signed by Garrote himself, numbered 237 of 500. "This one's probably worth at least a hundred bucks. Maybe more."

"Let's try the upper level," she suggested.

"I guess it's worth a shot," he said, but he didn't feel very optimistic about it.

He held the ladder while she went up, then he climbed up behind her. From here the spiral pattern in the floor tiles was more obvious. So was the black stain at its center. He inspected a shelf on the far left, thinking he recognized the spine of one of the books. "Hey. This book doesn't belong here."

Lilian approached him. "Which one?"

He pointed it out. One of the Garrote hardcovers had been misplaced among several scientific books, titles like *The Age of Spiritual Machines*, *Computational Intelligence* and *The Brain Makers*, along with a well-worn paperback of *The Terminal Man* by Michael Crichton. Ben grabbed the copy of *The House Feeds* but it wouldn't pull down from the shelf, it only tilted slightly, and a hidden mechanism made a sharp click.

"*Careful, dammit!*"

Garrote's voice came from behind them. They wheeled around and peered over the railing. Garrote stood over a disheveled man in a wooden chair. Ben gripped the railing, frightened for a moment, until he realized that this was not the Garrote they had seen elsewhere in the park.

"They're holograms," Lilian said. "You must have triggered a cutscene when you pulled on that book."

"Don't splash it on him," Garrote said to the woman standing beside him. She was dressed in a white suit jacket and skirt, with her back to them so they couldn't see her face. "It needs to appear as though I've poured it over myself."

The woman wore leather gloves and poured gasoline over the head of the unconscious man on the chair from a red plastic gas can. It soaked the man's straggly hair and spilled over his shoulders and chest, splashing on the floor around him. The unconscious man wore one of Garrote's cardigans but otherwise appeared filthy, as if the two of them had recently picked him up off the street.

"Care to do the honors?" Garrote said, holding up a single wooden match. It was the same blue-patterned matchbox as the one in the pocket of Ben's cargo shorts, the same brand the police had found tossed up against a bookshelf at the scene. If he'd been able to inspect it closer, he would have seen the bearded Chinese man on its face above the name Kitchen God Strike Anywhere Matches.

The woman refused Garrote's request with a brusque shake of her head and the writer began to smile. "Very well." He struck the match and let the flame flicker for a moment. It made his eyes twinkle. Then he tossed it onto the unconscious man.

Immediately the man burst into flames. Garrote laughed, the firelight giving his features a demonic look while the man in the chair screamed, his entire body engulfed by the blaze. Garrote's cardigan burned off the man's torso and the flesh began to peel off his bones. Forensics had found he hadn't been tied to the chair, yet he remained seated as if he was unable to get free, his back arching, melting fingers clawing at his thighs.

The woman had turned away. She was facing Ben and Lilian now but the flames at her back obscured her features. Behind her the dying man's stomach burst open and viscera spooled out over his lap, cooking between his legs. The woman's shoulders bunched as if she might be wincing or even retching.

"Oh, don't act so squeamish," Garrote scolded her. "You've done worse for your pound of flesh. Here, give me the gasoline."

He snatched the gas can from her and tossed it aside, where it landed near the matchbox. The police would later find both covered in Garrote's prints and conclude that Rex Garrote had burned himself alive—all but Detective Beadle, who thought something else entirely. Judging by what they'd just witnessed, the detective had been right.

Ben turned back to the science books, thinking they might be a clue. He'd read *The Terminal Man* on a Crichton binge one summer along with the *Jurassic Park* books and a handful of others. It was about a man who receives a surgical implant meant to counteract seizures by direct intervention in his brain. Before the implant the man believes machines will one day take over the world and after, he starts thinking he's becoming a computer himself. And in Rex Garrote's *The House Feeds*, a man's body and soul are slowly absorbed by the demonic Victorian home he lives in, Deaver House, which was based on this house.

It all made sense. But was it possible?

"Garrote is still alive," Ben said.

Before Lilian had a moment to process this information, the holograms flickered and twitched like a bad signal on an antique TV. The chair legs snapped under the man and his charred corpse dropped in slow motion. Suddenly the woman stepped briskly out of silhouette, out of sync with Garrote and the burning man, whose movements had drastically slowed. Ben recognized the woman's face, although the last time he had seen this person it hadn't been a woman.

It was the programmer standing there in a white jacket and pencil skirt. Harrison's face was unmistakable, with his shiny, balding head and those dark little rat eyes behind a pair of narrow, dandruff-flecked glasses, despite resting on the slim, hourglass figure of Rex Garrote's accomplice.

"*Kids! Thank God—I found you!*" the programmer cried. His voice had the same pre-echo quality as the ghost they'd helped leave its body outside the farmhouse. His voice stabilized as he approached the balcony below them. "If you want to get out of here you need to d-do exactly as I say, you g-g-got it?"

Lilian spoke. "You're still alive?"

"Just barely. I've inserted a digital avatar of myself into the program."

Ben said, "The reboot."

"Exactly," the programmer said. "But Garrote knows I'm in the system and he knows you're here, too. He'll get to us all soon, so we have to act quickly."

"How do we get out of here?" Lilian asked.

"You were on the right track with the books. I'll see if I can open the door for you the easy way."

The programmer squeezed his eyes shut, frowning in concentration. Behind him, Garrote's eyes began to slowly rise from the sight of the burning corpse, practically glowing with firelight, turning toward the man wearing his agent's body. The gleeful menace twisted into unadulterated rage, and Ben knew—he *knew*—Garrote wasn't a hologram, not any longer. The writer was here, in the room with them. And if Harrison didn't hurry, they would all be dead.

"He's here!" Ben cried.

"I'm doing the b-best I can. There's a lot of tech in this park, I just have to find the right algorithm to—*there!*"

At the far end of the balcony, the shelf popped out from the others with a click and groaned open, revealing a darkened hallway beyond.

"Quickly, get into the tunnel! You have to get to the servers. That's his only link with the system. I'll reach out to you as soon as I can but you have to—"

Garrote's eyes burned red and the woman's clothes burst into flames. The programmer screamed as the fabric peeled off his flesh. Ben looked away. He saw Lilian had already started across the balcony, heading for the secret door, and he chased after her.

They had to get to the servers. Garrote's consciousness was contained within it, exactly as he'd begun to suspect while looking at the books on the shelf. Somehow, Garrote had merged his consciousness with the computers, and those computers now controlled everything in Ghostland.

Pull the plug, sever Garrote's link to the park, to the ghosts, to his army.

Ben just hoped they could find the servers before Garrote caught them. Because this time, just like in his novels, the writer would leave no survivors.

LILIAN BURST forward with a surge of adrenaline as the passageway already began to rumble shut. Ben was steps behind her. In a blink of an eye Garrote was standing on the balcony above the ladder. Lilian hurried behind the bookcase. If Ben didn't reach her soon, he would miss the opportunity—the passage was almost shut. She held out a hand, fingers splayed, calling out for him to hurry. His breathing was heavy from the brief sprint along the balcony. She feared he wasn't going to make it.

Another blink and Garrote stood in the space Ben had just dashed away from, his eyes bright with malice, a dark smile frozen on his face.

Lilian backed into the dimly lit tunnel as Ben reached her and began squeezing his way through the narrow gap. He gasped, the back of the shelf pressing against his ribs, groaning as it tried to close with him inside the gap. She grabbed his hand and pulled, coaching him. "You can do it! Just a little bit more!"

Garrote was so close now. "*You... are... MINE!*" he roared. He snatched out and grabbed at Ben's T-shirt sleeve but his fingers slipped through the fabric and Ben managed to squeeze his head and shoulder through as the shelf drew closed with a click.

"Is he gone?" Ben said in the dark, gasping for breath. "Did we lose him?"

His exhaustion, his despair was palpable. She felt for his hands and squeezed them, hoping to pass some of her remaining strength on to him. "Let's just keep moving," she said. "I got a look down the tunnel before it closed. There's a lever at the far end. I bet it controls the door."

"Okay," Ben said. He took a deep breath and repeated himself. "Okay." He swallowed with difficulty.

She pumped his hands one last time for courage and let them go, then reached out into the darkness for the walls. The brick was cool, almost damp to the touch, the tunnel barely wider than her shoulders were broad. She moved forward, holding the walls, hoping she wouldn't trip, hoping Garrote wouldn't find them in the dark and drag them down into the blood-red water Allison had spoken of when she died.

"It's cold in here," she said. It was probably cold enough to see her own breath, if she could have seen anything at all. "Are we going down?"

"Huh?" Ben said behind her, still breathless.

"It feels like the tunnel's going down."

"I dunno." He paused, the only sound in the tunnel their footfalls, and the light swish of their hands against the cold stone walls. "Yeah, I think so."

The tunnel seemed to go on much longer than it had looked before the light shut out, leading down and down on a low decline. She wondered if it might take them right down to the basement. Her Converse scuffed on the floorboards. Her fingertips gently whispered against the walls. Her heart thrummed in her ears.

Suddenly she realized she could no longer hear Ben's labored breathing.

"*Ben?*"

"I'm here." He was only a step or two behind her.

Finally, she felt the cold metal lever and closed her fingers around it. Ben bumped into her from behind.

"S-sorry," he said.

"It's okay." She pulled the lever down. A sliver of light appeared in front of them, blindingly bright, in the shape of a door. Brick ground against brick as the passage opened on a walk-in freezer, the walls lined with shelves stacked with boxes of all sizes.

That explains the cold. "I guess we're not in the basement," she said.

"N-no. Probably not."

Goosebumps sprung up on her arms and her breath plumed out before her. The passage hissed closed behind them, a neat rectangle of insulated metal no one would suspect was a passage from inside the fridge. She reached the exit and slammed a shoulder into the door, jerking down on the handle. It wouldn't open. "We're stuck!" she said, hugging herself as she shivered.

"There's a pin," Ben said. He reached past her and pulled the metal pin out of the handle.

She tried the door again and it opened easily. Warmth bloomed on her cheeks and arms as she stepped into the kitchen. Her first breath formed a cloud of vapor as she slammed the fridge door behind her. Her next breath less so. The third wasn't visible at all. The light hairs on her forearms settled and the flesh on her forearms smoothed out, the goosebumps disappearing.

They stood in a large kitchen of pale tile and shimmering steel. She turned to Ben. He was breathing deeply, slowly, in through the nose, out through the mouth.

"You okay?"

"Uh-huh," he said eagerly.

A shrill scream sounded nearby, startling them both. Lilian traced the sound to the kitchen door, where a pretty young black woman with her hair in a tight bun burst through, dressed in a vintage maid's uniform with a white ruffled collar and apron. As the maid glanced over her shoulder she tripped, falling to her knees against the foot of the counter island. She knelt beside it, cowering, watching the door as she blindly felt the countertop for something to defend herself with.

A shotgun blast punched a ragged hole through the door, thrusting it inward.

The maid let out an involuntary scream, stifling it with the hand not roaming the countertop. Her employer stepped in: a fat, sweaty, rosy-cheeked man with a neat little waxed mustache, dressed in a cinched tuxedo with the buttons at his waist ready to burst. He expelled an empty shell from the shotgun in his hands and took aim.

"Mista Hedgewood," the maid said. Her slender fingers touched the serrated edge of a bread knife and her eyes widened. "This ol' house got its claws in you."

"Hedgewood," Ben said in a revelatory tone, as if the name meant something to him.

For a moment Lilian worried the ghosts would react to his voice, but they seemed either disinterested or unable to hear him. The man in the tuxedo—Mr. Hedgewood—considered the maid's proclamation for a moment, but he pulled the trigger anyhow. The maid winced against the impending pain... but the shotgun didn't fire, the trigger only clicked dryly.

"Blast it!" he cried.

He turned the barrel up to his eye and peered down into it. In the instant his attention left her, the maid closed her fingers tightly around the shaft of the knife and jabbed it into the man's stomach. His mouth opened in a cherry-red O and the shotgun went off in his hands. The upper half of his head exploded in a wet, chunky mess and the large man fell forward, toppling over her.

Drenched in his blood, she dropped the knife, which clattered on the tiles as she hurled the body from herself in horror and disgust. Hedgewood slumped to the floor, the ragged territory of flesh and gristle and bone above his neck pouring blood onto the white tile.

Her face painted in the dead man's blood, the maid's wide-eyed gaze settled on Lilian and Ben as if she could see them, more than a hundred years into the future. "He's coming," she said. But it was Harrison's voice that came from her lips. "I'll unlock the door under the stairs but I can't hold him off much longer. You have to move, kids."

Hedgewood's blood-slicked palms slapped down on the tiles beside the woman and he began to push himself to his feet. The corpse staggered blindly, his exposed throat gurgling as gore slopped out of the lower half of his head like an overfilled soup bowl. The tongue flopped with a horrible glottal sound as his fingers reached blindly for the knife in his stomach.

"Mista Hedgewood," the maid said, fear in her voice as she backed away from the living corpse on her hands and feet. "You cain't be alive. You ain't got a head..."

Suddenly all the pots and pans on the hanging rack started swinging, clanging hollowly against each other, and the stove burners all turned on at once, burning with large blue flames, and the cupboards and drawers slammed open and shut, open and shut, their contents clattering.

"Let's go!" Lilian shouted over the noise. She grabbed Ben's hand. His fingers were still so cold, as cold as the grave.

They ran for the door as utensils flung out from the drawers, hurtling toward them. Lilian ducked from a large shiny knife which sunk into the wall ahead of them with a doorstop twang, knocking an antique brass clock off the wall. Cutlery and condiments struck the walls and clattered on the tiles at their feet, clouds of flour, scattered cereals, cracked eggs. As they hurried past the headless corpse, he swung at them with the knife, the swish of the blade audible over the racket.

Ben shouldered through the door into the next room. Lilian hurried in behind him, pushing it shut and twisting the key in the lock. Her heart pounding, she turned to look at the room they had entered. She gasped at the sight of yet another horror.

The dining table was set for a fancy dinner. The remains of a large turkey and various side dishes moldered, buzzing with flies and the Rice Krispies crackle of a thousand crawling maggots. Around the table sat a young family in Victorian dress: a mother, a father and three children of varying ages, their corpses purplish and bloated, a foamy pinkish crust on their chins and dribbling on their napkins and bibs.

The father's head rose, lolling drunkenly. "He that eateth, eateth to the Lord," he gurgled, blood and foam pouring from his lips. "For he giveth God thanks."

"Amen," his family gurgled in unison, then they began shoveling rotten morsels crawling with insects into their mouths.

As this macabre scene played out a deep, rumbling horn bleated from beyond the French doors to the foyer, so loud it rattled the glass in its frames.

"What was that?" Ben whispered, terror in his voice.

The sound had frightened Lilian too, rattling her heart and setting her nerves on edge. Behind them, something pounded against the kitchen door, a metallic scrape tracing its way down the wood.

Hedgewood, the headless man, had found them. They would have to risk running into whatever had made that terrible sound out in the foyer to get to the basement stairs.

"Come on," she said.

They sidled past the ravenous ghosts busily devouring their squirming, larvae-infested last meal and stopped in front of the French doors, attempting to peer through the etched glass. The foyer looked deserted. Whatever had made the strange trumpeting sound was gone.

Behind them the kitchen door splintered around the handle and the headless man shambled in, swinging the knife. Lilian and Ben both grabbed a handle. Ben gave her a worrisome look—obviously as wary of the sound they'd heard as she was—and together they drew the French doors open. Once through, they slammed them shut. They'd ended up at the far end of the foyer, near the front door.

"Look!" Ben shouted, pointing toward the basement door. The light on the lock mechanism was green. Harrison had opened it, as promised.

As they hurried across the foyer the loud, rumbling trumpeting came again. In the instant she realized where it was coming from, the basement door swung open violently, crashing against the base of the stairs.

For a long, silent moment nothing happened. They stared into the darkened opening, exhausted and afraid, awaiting the inevitable. Finally, a massive, impossible creature emerged from the darkness, so tall it had to duck to get through the doorway.

"Oh no," Ben gasped. "It's *him*."

Lilian didn't know who *he* was and she didn't want to find out.

She only knew the hideous, inhuman giant stood between them and escape. And with the headless Hedgewood on their heels, Garrote awaiting them in the library and all the doors and windows barred, they had finally reached the end.

There was nowhere left to run.

THE BEHEMOTH

BEN COULDN'T MOVE, staring at the monstrous Final Boss creature that had lumbered into the foyer. He'd heard rumors about this ghost from the few survivors who'd dared break into Garrote House at night after the writer's death. Chat groups on ghost hunting and the paranormal called him the Sculptor, the Behemoth or—less frequently—Picasso's Monster. It was said to be an amalgam of body parts from the Odell family, torn apart and fused together. Looking at the thing towering over them Ben could see this rumor was true.

From the shoulders down, the Behemoth appeared to be a normal man: the famous sculptor, Clayton Odell, whose body of work had been compared to H.R. Giger's biomechanical paintings. His dark skin and muscular legs were partly hidden beneath a long, stained leather apron speckled with solder burns. In one hand he held a welding torch with a sharp blue flame. In the other, his thick, gloved fingers gripped a heavy scrap of burnished metal that looked like it might have belonged to one of his macabre sculptures displayed on either side of the fireplace. Like the torch, it could easily be used as a weapon.

And just like chat group gossip had stated, Clayton had no visible head of his own. Instead, the head and torso of his pale, freckled wife Laura was perched on his shoulders, severed at the waist and fastened to the sculptor's shoulders by metal straps. Blood from her wounds had oozed down the man's shoulders and spattered his apron. Her long, sleek black hair hung over her face, her head lolling lifelessly, adorned with a mask made of welded brass with a cone jutting outward like a megaphone that held her jaws impossibly wide. It reminded Ben of the medieval torture device called the "scold's bridle," or "witch's bridle," although this seemed meant to amplify the woman's voice rather than throttle it. Where her breasts should be, twin toddlers had been crudely sewn to her chest from their bellies up. The gray-skinned babies drooled, their eyes sewn shut, their chubby fingers grabbing mindlessly in front of them. Reaching for what, Ben had no idea.

The dead woman's head cocked at an angle, and her green-eyed gaze fell upon Ben and Lilian. A hand rose zombie-like to point at them, and she screeched into the horn. Her cry caused the bone-chilling bleat they'd heard earlier to echo coldly throughout the empty house.

Clayton's bare feet shuffled at the sound, turning his wife and children toward them. The Behemoth, Picasso's Monster, raised the welding torch in its gloved hand and began lurching forward, shaking the floorboards and rattling the lamps on the bannister. As if it were solid. As if it were real.

Ben stood there, still unable to move, as fascinated as he was frightened. The floors must have been rigged to tremble with its footsteps, like the seats in a 4D ride, to make the monster's presence even more menacing. It worked. He was so frightened he had to force himself to take a cautious step toward the stairs.

The floorboard creaked under his foot. Again, he froze. The Behemoth's head snapped toward him. Its muscular left arm swung out with the giant scrap of metal. Its sharp edge missed him by several feet. But if they didn't get out of here quickly, it would cut them to shreds.

A gurgling groan arose from the dining hall. Ben turned to see Hedgewood's headless corpse shamble out into the foyer with the knife. His lower jaw had come unhinged and flopped wetly against the once-white shirt and black tie.

There was no way out. The stun gun seemed useless, a toy, like going up against a gun holding a sharpened stick.

Lilian shouted, "Now would be a really good time to show up, Harrison!"

The programmer didn't respond and the Behemoth plodded forward, reacting to her voice. The headless corpse stabbed the air with the knife, his fat tongue flapping in the blood pooled in the basin of his shattered skull, like a dog's lapping in its water dish.

The ghosts were closing in on them. One or both would reach them soon.

Lilian grabbed Ben's hand and pulled him away. The Behemoth's reach with its jagged blade could easily cut off their heads if they attempted to run for the basement door. She led them to the stairs. Ben shook from his stupor as he ran, and bounded up by her side.

They hit the second-floor landing at full stride. Ben dashed through the first doorway he saw, stopping only when he reached a long, dim corridor. Paintings with display lamps above them lined the walls, hung between doors to their left and on either side of the balcony overlooking the foyer to their right. A pastoral setting, a gothic castle at night, a stiff Victorian portrait of the Hedgewood family. Ben thought he saw movement in the portrait from the corner of his eye but when he turned to look the subjects of the painting remained still.

He peered out through the balcony door. The ghosts hadn't followed them. Hedgewood and the Behemoth appeared to be in the midst of a blindfolded knife fight.

They were safe for the moment. Which was fortunate. Lilian was starting to look tired. Like him, she'd been pushed beyond exhaustion. And now they'd gone out of their way again, making their destination even further away. He thought about how far they'd come to get here, how many friends they had lost along the way, and the thought gave him a desperate sinking feeling in his stomach.

He had to tell her, before it was too late.

"Do you have any idea which way to go?" he asked instead, and cursed himself silently for losing his nerve.

"I don't know," she said. "We can't go back down there, that's all I know."

"No," he agreed. "I was thinking... a place this big might have a second stairwell. A lot of these old houses had stairs from the servants' quarters to the kitchen."

"Makes sense. But where?"

"I dunno. But I figure if we put our heads together, we're sure to find it."

"Let's hope so," Lilian said, without much optimism in her voice.

Ben smiled wanly as she started ahead. He watched her walk three swift paces before he finally managed to call her name, all but choking it out.

She turned back, cocking her head quizzically. "What?"

His tongue held back the words, refusing to let them go.

Tell her, dammit!

She wouldn't like what he had to say. He knew that. She would argue.

After all they'd been through today, she might even cry. But it needed to be said. He took a deep breath, exhaled slowly, and spoke. "I just wanted to say, if anything happens to me—"

"Nothing's going to happen to you."

"But if it does," he said, resisting the urge to backtrack, to let her convince him he was wrong to even suggest it. "You have to keep moving, Lilian. Don't stop, not even for a second. Just leave me behind."

She gave him a look of uncertainty, bordering on anger. "I'm not going to *leave* you, Ben."

"I *want* you to. Please. If... if he gets into me, if I start acting strange—"

"How would he get into you, Ben? What do you even *mean*?"

"Just listen to me for a second, goddammit! Jesus!"

She huffed, her shoulders sagging in the baggy silver fabric. "Okay..."

"You were always the best of us," he said. "That's why you need to live through this, Lilian. I won't have the same opportunities you will. You're smarter, you're personable. Do you remember that Halloween—must've been sixth grade—when we fought off all those big kids who were egging the first graders?"

She nodded, evidently curious where he was going with this.

"You stood right up to them. You were afraid, I could see it in your face—heck, you were shaking so bad it looked like you just climbed out of the water down at the Hole—but you didn't back down until you convinced them they were wrong and they all went home with guilty looks."

"Yeah," she said, smiling slightly at the memory.

"That's when I knew you were bigger than this place," he said. "I knew you'd go on to do big things. *Great* things. You could be anything you want to be. I wish I had that."

"You're smart," she said.

He smiled patiently. "I know. But I know my limitations. I could never go very far. There's always a chance I could die tomorrow—"

"Either of us could die tomorrow."

"Lilian," he said, approaching her. "Be serious, okay? I should be dead already. That day, when this... when this fucking house came through town, Garrote got to me. He *did* something to me, got inside my head—*something*. I don't know what, but I had to come here today. I knew I had to stop him. And I dragged you here—"

"You didn't—"

"I *did*, Lilian. You wouldn't be here if not for me. And if you hadn't, there's no way I would have survived this long. I would've been dead by lunchtime. You've carried me this far—"

"We carried each other!"

"Look, we can keep arguing about this until Garrote shows up to kill us both or you can just promise me right now that you'll keep moving—"

Lilian stepped up to him and brought a hand to his face. For a moment he thought she might slap him. Instead she grabbed the back of his head and pulled him to her, surprising the hell out of him when she pressed her lips to his. He kept his eyes open in disbelief. Lilian had closed hers. It was his first for-real kiss, and her lips were warm and moist with life, unlike the cold, dead lips of the nun—

Don't think about her, dammit! You've thought about this for years and now you're wasting it thinking about that—

—and then it was over. She stepped back and looked down slightly into his eyes, a thin rope of saliva still linking their lips. It broke apart and she cocked her head again, studying his face.

"I shouldn't have done that," she said suddenly. "I don't know what I was thinking."

"No," he said. "It was—it was good."

"You're not just saying that? I mean, it was weird though, right?"

"Yeah, of course. I mean no. It was—I liked it very much," he said, feeling the heat of embarrassment rise in his cheeks again. He didn't know what to say that wouldn't sound stupid or accidentally hurtful. It was good and weird and crazy and all of those things. It was everything he'd thought it would be and more. If he could only form the words to express how much it had meant to him. But the skill was beyond him. He looked down at his shoes.

"I liked it too," she said. She touched his face, making him look up at her. "You're not going to die on me. Okay? We're gonna make it through this. *Both* of us."

She held his gaze until he agreed. It didn't feel like a lie or a compromise. Never in his life had he wanted to live as badly as he did now. Because Lilian had liked it too.

"Now come on," she said, taking his hand. "Let's go check out some ghosts one last time."

Ben followed down the hall alongside her, his heart full and brimming with confidence for the first time since it stopped beating four years earlier.

OLD BONES

LILIAN HAD SEEN the look of hopelessness in Ben's eyes. That lost spark. She'd only wanted to give him something to live for, something to keep fighting for. But now the kiss was between them and she couldn't take it back. She wasn't even sure she wanted to. There had never been anything physical between them. When they were young, they'd been more like a brother and sister to each other than friends, and she would have laughed at the idea of them ever being boyfriend and girlfriend. He was nothing like the boys she was normally attracted to... and yet, despite the sharp smell of his fear sweat in her nostrils and the fumbling quality of the kiss itself, she couldn't deny the intense rush of feelings she'd felt, almost as if—

"Where should we start?" came a man's disembodied voice, startling her from her thoughts.

"In the master bedroom," a woman replied. Her voice sounded familiar. "Where *he* started."

"What now?" Ben groaned.

Lilian shook her head, looking around warily. She couldn't deny not feeling at least a little grateful for the interruption. High heels clacked down the hardwood floors, moving away from them. The sound stopped abruptly and the second door from the end of the hallway opened with a distinct creak.

"Come on," Lilian said, heading toward it. Ben followed her. They approached the room cautiously and peered in through the doorway.

Sara Jane Amblin stood just inside the master bedroom, dressed in a jacket and pencil skirt combo similar to what she'd worn in the control room. A slightly taller man stood beside her, dressed business casual with a blazer and Polo shirt, pressed suit pants and sneakers. The two of them wore large headsets—more like clunky, old VR tech than the sleek AR glasses Lilian and Ben wore—and were looking down at the large four-poster bed. Through a gauzy white curtain, Lilian was able to see a black man with short-cropped hair and a woman almost ghostly pale with long dark hair, both of them sleeping peacefully. There were photographs of the couple on the dresser. On the wall beside the bed hung a painting of sheep grazing in a pasture. A small brass sculpture stood in the corner of the room, near the door to an en suite bathroom.

"The master suite," Sara Jane said. "Rex Garrote slept here. And Laura and Clayton Odell, the sculptor and his wife, before him. And long before any of them lived in this house it belonged to a wealthy San Francisco shipping tycoon named Oliver Hedgewood[1]." She turned to the man. "Your great-grandfather."

The man nodded thoughtfully and approached the bed. His hair was wavy blond, graying at the sides. He had high, sculpted cheekbones and full red lips. His forehead creased in a frown as he fingered the gauzy curtains, pulling them back to watch the couple sleep. In the bed, Clayton Odell murmured and rolled to his side, away from his wife, facing the door.

The man standing over the bed adjusted his headset with an annoyed grunt.

"The headsets are prototypes," the inventor assured him. "The final product will be much smaller. We've partnered with a major sunglasses manufacturer to ensure they're more comfortable to wear, as well."

"That's good to hear. And this is where Odell...?" He seemed not to be able to finish the thought and so Sara Jane nodded.

"We've looped the final half hour of the Odell family's lives. In a moment, he'll wake," she said. "According to the biography penned by his niece, he'd begun sleepwalking shortly after the family moved in. She'd spent a summer with her aunt and uncle when the twins were born. She was the only survivor when—"

"—Clayton Odell sleepwalked to his studio, picked up the reciprocating saw and cut his wife in half at the waist while the poor woman screamed for him to stop, clawing at his face."

"You know your history," Sara Jane said, seemingly impressed. "The master of the house, your great-grandfather Oliver, he got off easy compared to the Odell family. You've seen the Behemoth? What some call Picasso's Monster?"

"The Sculptor," the man said with a grave nod. "I'd seen it once or twice, when I was young, after my father bought the house back." He shuddered and let the curtains fall back against the bedframe. "Out of the corner of my eye, of course. Awful thing. Gave me nightmares."

"As one of our star attractions he's sure to give our customers nightmares, as well. Mr. Garrote was insistent the Behemoth be front and center once the park is up and run—"

Clayton Odell shot awake with a gasp.

"Step right up, ladies and gentlemen," Sara Jane said, adjusting her headset as her lips perked up in a smile. "The show is about to begin..."

Clayton swung his feet out from under the covers and onto the floor with a vacant gaze, and all of the holograms evaporated, leaving Ben and Lilian alone in the bedroom.

"Hedgewood," Ben said. "He's the silent partner. The family who built this house, they're the ones who helped fund Sara Jane's research and build Ghostland."

"You think they knew what Garrote planned to do?" Lilian asked. "You think they were in on it?"

"I don't know. But they knew how evil this house was." He corrected himself: "*Is*. They knew that and they still charged people forty bucks a ticket to walk through here. If they were in on his plan or not—"

The room next door opened, as if on cue. Ben and Lilian crept over and looked into the nursery. There were two large antique cribs which looked like old-fashioned sleighs, painted white, with floral designs and frilly white skirts. A woman with skin the color of soap and a long black dress sat in a red velvet chair between the two cribs, cradling the remains of a baby swaddled in a blanket with a bonnet tied around its shriveled, gray-green face. The governess rocked the dead child on her knee, making gentle shushing sounds as flies circled around their heads. After a moment, she looked up at Ben and Lilian standing side by side in the doorway.

"Shh," she whispered, placing a finger against her lips. "The little one is sleeping."

Ben shuddered and drew the door closed.

They turned to face the final door at the end of the hall, perpendicular to the others. Its keyhole bore multiple scratches, the brass door handle battered and blackened from age. For a moment Lilian worried they might need a key, but before she reached the handle the latch clicked and the door creaked open. She'd hoped to find a staircase leading down. Instead, it led to another hallway.

They stepped through into the corridor. The last of the evening sun shone through dusty lace curtains draped over cracked windows yellow with grime, the glass warped from age. The opposing wall had been decorated with the bare minimum of wall sconce lighting, no paintings or photographs, and only two doors: one in the middle of the long hall and one at the far end, facing them.

Lilian waited for Ben to catch up before moving onward, hugging the wall opposite the windows. She half expected something to come crashing through the glass and attack them and it seemed Ben felt the same. He held the stun gun out in front of him, finger on the trigger, shooting jerky looks at the windows every so often, as distant cries came from outside.

"Survivors?" he asked, sounding hopeful.

"It's just seagulls," she said.

"Oh." He looked just as disappointed as she felt.

Thinking about the two of them alone, against all odds, she said, "I'm sorry I ghosted you for so long. That wasn't fair." She was feeling fatalistic again. The urge to tell him, to admit her mistake in case neither of them made it out of here alive loomed over her like a presence, a ghost of their past. Ben had already made his confession. It was her turn.

"It's okay," he said.

She shook her head. "It's not okay. Death scares the hell out of me, Ben. It always has. That's why I started watching horror movies, playing all those games with you. It was a safe outlet. That's what Allison told me. I could look at death from a distance." She felt tears coming. "That's why I couldn't be friends with you anymore. Because when you died it made death real. It was all I could think about. Everything I did, death was right behind the corner waiting for me. But then I realized—Allison *made me* realize—all this time I've spent hiding from you I haven't just been hiding from death, I've been hiding from *life*. I guess I always kind of knew it but coming here with you proved it to me. So let's live, okay? For everyone who died here." She took Ben's hands, the bulky gloves an almost agonizing barrier between them. She craved contact, to feel the warmth of skin on skin. "Let's *live*," she said again, smiling through her tears.

Ben swallowed hard and nodded, both of them crying now. He wiped his eyes with the heels of his palms. Lilian let them fall.

Garrote's voice startled them both. "Careful, you idiot," he said, his words slightly slurred, coming from the nearby doorway. "I still need my damned tongue."

They peered into the room, saw Garrote seated on a toilet and a man with his back turned to them, dressed in a blue Polo shirt, reaching into Garrote's mouth with a pair of pliers. Garrote's right hand was wrapped in bandages soaked through with blood. A half-drunk bottle of scotch labelled Macallan 10 Years Old stood beside the toilet. Garrote winced as the man pulled a tooth from his open mouth and dropped it into the blood-spattered sink, where it clinked against the porcelain.

"Who's that with the pliers?" Ben asked, moving to try and get a view of the man's face.

Lilian looked in the mirror to find out. His face was just out of sight but in the corner of the mirror she saw another man, one unseen by Ben and the others. He lay in a tub full of black water, his eyes white globes of fear staring directly at her from within a pitch-black face pitted and scalded, his hair melted away in clumps, his shoulders and arms and chest a topographical map of pustulating sores and burns speckled with small white feathers.

Lilian backed away from the room, the terrified eyes of the tarred and feathered man etched onto her retinas when she blinked. She squeezed her eyes until the image faded.

Plink! Another tooth dropped onto the porcelain.

"Are you okay?"

"I saw..." she began. She just swallowed hard and shook her head.

"What was it?"

"*Don't look*," she warned, and Ben stepped back from the bathroom door without question.

Only one way left to them now. The door at the end of the hall was painted black as death. Lilian felt like they'd been pushed this way, lead here, everything in this house poking and prodding them toward this last door, this final threshold.

She couldn't help but feel that whatever lay beyond that door, it was something she wouldn't want to see.

"Are you thinking what I'm thinking?" Ben asked.

She nodded, no need to see his face to know it was true. She could hear it in his voice.

And before they could make a decision to keep going or turn back the way they'd come the house again decided for them as the hallway burst into flames. The wallpaper bubbled and peeled, the curtains blackened and curled, scraps of burning fabric fluttering toward the ceiling, where flames rolled toward them in red-hot waves, and at the end of the hall a woman screamed, burning from head to toe, flailing her arms and running zigzag toward them, heaving against the walls and pinwheeling away as her clothes peeled from her charring flesh.

Lilian didn't waste another second. She turned, yanked the black door open and leaped blindly into the room, Ben at her heels. The fiery woman fell in a scorched heap at the foot of the door, her flesh sloughing off in sheets and melting into the Persian rug like gobs of candle wax, setting its fringe on fire. With her bones exposed the woman still writhed, her jaw and all of her teeth visible as she screamed, begging for her agony to end. Even after the burning woman's screams died her carcass kept moving, squirming and bubbling, reminding Lilian of the black snake fireworks Ben used to light on the Fourth of July, the ones that grew and curled over themselves as they turned to ash.

She slammed the door and leaned against it, finally turning to look at their surroundings.

THE HOUSE SPEAKS

GARROTE'S OFFICE WAS large, with polished wood floors and a high ceiling. A brass-mounted globe stood at its center and a lion's head hung above a fireplace. An armchair had been placed on one side of the hearth, a samurai helmet in a glass display case on the other. At the far end of the room stood a large mahogany desk, inlaid bookshelves on the wall behind it. Images in frames dotted the walls elsewhere, some large, others small. Mostly book covers behind glass, the few others photographs of Garrote with famous people, with his Army platoon, with his books.

Through the windows, Lilian could see what was likely the top of the wall surrounding the park. For a reckless moment she thought that if they could leap across the chasm between the window and wall they could escape, but it looked far too wide. They'd need something to cross it. The fall would be too high to survive without severely broken bones.

"I've seen this room," she said.

"This is where I saw him that day." Ben pointed to the window in the corner. "Standing in that window."

Lilian shivered, remembering the enormous shadow that had flitted past the opened doorway the day Garrote House rolled through town. She backed away from the black door, wondering if it had been the remains of the Odells —what the Ghost Brothers had called "the Behemoth"—that she'd seen through the window. She wondered, if she'd been the one to see Garrote that day instead of Ben, would she have been the one whose heart stopped beating? Would she have been forced to live under Garrote's shadow the past four years, wondering *Why me?* Forever wondering if Garrote would come back for her, to stop her heart for good? Had it all come down to bad luck? Or had Garrote *wanted* Ben to see him? For the thousandth time she cursed herself for ever having told Ben to look out the window that day, for having shown him this damned house.

Unaware of her guilt, Ben crossed to the globe. "This is a Lenox Globe," he said.

"Isn't that what you and the detective were nerding out about this morning?"

He set it spinning and grinned back at her, the globe a whirling blur of greens, brown and blue at his hip. "It's weird," he said. "You're the one who's good with directions but I'm the one who loves maps." He slapped a palm onto the globe to stop it from spinning. Lilian imagined all the tiny people on the Eastern seaboard his hand would have crushed.

How many people died by Garrote's hand today? she wondered. *Hundreds? Thousands? Is anyone still alive out there? Are we the last?*

"All right, all right, wait your damn turn!"

Lilian startled at the sound of Garrote's voice, while Ben lifted his hand from the globe and aimed the stun gun.

The writer's hologram sat hunched over the typewriter, his hair a mess, dark patches under his eyes like shriveled teabags, a smoldering cigarette clamped between his lips. Balled-up sheets of paper were strewn everywhere on the desk, the floor. There was a half-empty bottle of Macallan at Garrote's right elbow, alongside an ashtray full of cigarette butts. Several more bottles poked out of the trash bin beside the desk. The writer typed in a flurry, then paused, listening.

"Yes, I can hear you, dammit!" he shouted. "I'm typing as fast as I can!"

"Who's he talking to?" Lilian asked.

"The ghosts," Ben said. "He's talking to the ghosts of the house. The ones who lived here before him."

As he spoke figures appeared half-formed around Garrote, more like true apparitions than the holograms they had seen elsewhere in the house. She recognized the Odells and their curly-haired twins from the photographs she'd seen—but also as parts of the monster called the Behemoth. Beside them stood a pasty-faced butler, holding his arms behind his back with his chest heaved outward, and the maid they had seen in the kitchen, who was now whispering in Garrote's ear. Then a pair of builders with handlebar mustaches, dressed in dirty overalls and flat caps. One man was studded with nails. The other had an iron rod jutting from the top of his skull. More ghosts stood behind them, huddled in a tight group between Garrote and the far wall: a Native American shaman, his rich brown skin painted from the waist up; the governess with the rotted child in her arms; the tarred and feathered man, the whites of his eyes stark against the shimmering black tar on his skin; the burning woman, her embers still smoldering; three bored-looking teenagers from the '80s, each missing parts of their face or an arm or bleeding through their vintage T-shirts.

Garrote took a long drag on his cigarette, nodding as the maid spoke into his ear. On his right Hedgewood began flapping his tongue, spattering his cummerbund with gore.

"Goddammit, Ollie, can't you see I'm listening to Lutessa? Wait your fucking turn!"

The maid finished speaking and Garrote began again to type. "These old bones," he muttered, smiling as his fingers clattered over the keys. "These old bones sure do tell the tale."

The long ash fell off his cigarette onto the blotter. He brushed it away absently and kept typing. Then they disappeared.

In the silence that followed, Ben spoke. "*A Roller-Coaster Ride Thru Hell*, Garrote's first novel, was super-successful. He made a ton of money from the paperback and movie rights, even though a movie never got made. But after that he wrote two critical and commercial bombs, *Shōki* and *Blood Red*." Ben shrugged, scratching his chest absently, over his heart. "They're okay books, not his best. He thought getting a change of scenery might help his writing, so he bought this house, on the other side of the country from where he grew up in Connecticut. After that he wrote one hit after another, starting with *The House Feeds*, about a writer possessed by the ghosts in his house. It's basically *The Shining* without all the psychic stuff. After that some fans came up with this crazy theory that the story was autobiographical, that the ghosts in this house had actually told him what to write."

"I guess it's true," Lilian said.

"I guess so." Ben looked around the room in thought. "You know, in *The House Feeds* there was a trapdoor underneath the main character's desk, leading down to these creepy old catacombs under the house."

"How did it open?"

"I haven't read it in a long time." He broke away from her and approached the desk. "The globe is too obvious. So is the lion's head."

"A book, maybe?" she suggested.

He shook his head. "He already did that in the library. Garrote hates to repeat himself. He says self-cannibalization is for lesser writers than him."

Lilian followed Ben to the desk. All of the clutter from the holographic flashback was gone, leaving only the old black Remington typewriter on a green leather blotter, a small gargoyle facing the desk chair with a clawed finger pressed to its smiling lips as if to hold back a secret, a mug filled with pens, and an antique ink quill set.

The chair squeaked as Ben sat in it. He picked up the gargoyle and examined it briefly before setting it back down. He hovered his fingers over the keys and mimed typing for a moment, getting into the zone. Then he rolled the chair back and peered under the desk.

"There's a wire under here," he said, sitting back up.

Inspiration struck and he raised the typewriter off the desk, revealing a thick black cable that trailed from its base to a hole in the desk. "The typewriter! Of course!"

Lilian rolled her eyes. "Of course."

He picked up a sheet of paper, rolled it through and began typing. For a machine that called itself "Noiseless"—according to the name printed across the top—it sure made a racket. The words came easily, as if Ben himself were possessed. When he'd finished Lilian looked over his shoulder and read it aloud.

"'The last living ghost looked over the burning ruins of the world he'd conquered and thought, *This land belongs to us now—a world of the dead, a world full of ghosts*.'" She cocked her head at Ben. "Did you just make that up?"

He shook his head, frowning at the page. "It's the first line from *Shōki*. A flashforward to the end of the book. I thought maybe it would trigger something, like a password."

"Passwords are usually one word," Lilian said. "What if it's just 'Shōki'?"

"Maybe."

He typed the word. Nothing happened. "There's no accents on the keys," he said.

"I've got an idea." She leaned over him and tore the page from the typewriter, then started pulling out the ribbon, getting ink on her fingers.

"What are you doing?"

She pulled out enough until what Ben had typed was no longer visible. Then she saw what she was hoping to find. "If this opens the way down someone must have used it before," she said. "Look."

Ben peered over her shoulder. He saw the same thing she did, the code *Shoki237*, repeated several times. "Like the signed hardcover in the library," he said. "I guess I should've thought of that."

Lilian shrugged. "Well, you didn't, so kiss my score and get out of the way, chump." She made to sit down beside him on the narrow seat. Ben stood up quickly, looking anxious. She ignored his embarrassment and carefully typed out the password.

As she struck the 7 key a sharp click came from beneath the desk. She leaped out of the chair as a section of the floor rose several inches then began rolling away from them, exposing a dark chasm below their feet. The chair was swallowed up by the expanding hole. It struck something hard, then something else, again and again, the crashes moving further from them, down and away. It sounded like stairs.

"Whoa," Ben said. "Cool."

Lilian squinted down into the dark passage. "Not much light down there."

Ben reached into the leg pocket of his cargo shorts and plucked out a box of matches. The box had a blue pattern and the name Kitchen God. *Weird name*, she thought.

He nodded and struck the first match, then crouched, holding it over the stairwell. The warm yellow glow illuminated dark wainscoting and candles in holders at intervals along the walls. The smell that arose from within was dank and musty, the stench of age and rot.

Lilian wondered when this passage had been opened last. Someone must have used it at some point in the past four years. The great gobs of melted wax on the candles indicated the electricity had never been updated in this part of the house. But the password had been used multiple times recently, judging by the imprints on the typewriter ribbon.

The floor began to shake, as if the trap door was already closing. Ben had already descended the stairwell up to his shoulders. He gripped the edge of the passage, eyes wide in fear, trying to look around the foot of the desk. "What's happening?"

Lilian didn't know. The passage wasn't closing, not yet. She couldn't think what could be shaking the floor, unless—

Out in the hallway a deep, rumbling horn blared, echoing through this house of the dead. Her skin prickled with fear.

The Behemoth had found them.

THE RED WATER

T*HIS FEELS LIKE a trap,* Lilian thought as she hurried down the steps to Ben's side.

The smell was even worse down here, like when her dad had pulled up the floorboards at the old house on the Duck Bill, and the stink of dozens of dead mice and dried feces had wafted up from the foundation. Ben struck a match against the box. Its sharp burnt sulfur smell smothered the stench as he held it against the wick of the nearest candle, revealing a long stairwell that descended into the darkness beyond a narrow landing and a door below them.

The Behemoth's horn bellowed again, ringing off the walls above them. It was closer now, hunting them through the house.

Lilian peered out at the slash of lowering sunlight as the floor rumbled back into place. With the keeper suit on, the ghosts couldn't harm her. But she couldn't protect Ben for long. She hurried down to the first landing, jerked the handle on the Maglock. The door wouldn't budge.

"That's probably the first-floor entrance," Ben said. He struck another match and lit the next candle. The stairs descended another ten feet, stopping at another door, where Garrote's desk chair lay on its side.

The Behemoth's horn blasted down into the passage. They looked up, gripping each other. Laura Odell's dead eyes peered down through the narrowing gap just as the entrance grumbled closed. Lilian and Ben huddled together in the darkened stairwell, the only light from the guttering candles.

Could the Behemoth follow them through the trapdoor? Lilian wasn't about to wait around and find out.

She hurried to the bottom of the stairwell, stepped over the desk chair and twisted the doorknob. Beyond was another long tunnel, a string of emergency lighting burning along the ceiling, giving the tunnel an eerie red glow. Bundles of thick cables came out of the wall near the ceiling and trailed down the hall below the lights.

They stepped into the tunnel. Ben closed the door behind him.

"I think we lost it," he said.

"Don't jinx it."

They followed the cables, their pale faces red from the lights. At the far end the lights trailed down another corridor to the right. When they reached it, Lilian began to hear a hum that rose in volume as they turned the corner, a hum she seemed to feel in her teeth.

"What's that sound?" she asked.

"Sounds like computer fans," Ben said. "We must be getting close to the servers."

"Why would they put the servers so far away from the control room? Does that make sense?"

"I was thinking that too. All these cables, hundreds of feet long. Seems like a waste of time and money."

Another door lay ahead, with a Maglock like the others. A sign beside the opened door read, LEVEL 3 EMPLOYEES ONLY BEYOND THIS POINT! Lilian assumed the programmer must have unlocked it for them, and wondered if he'd be able to open the hatch when the time came. That would be too convenient—movie critics would call it *deus ex machina*. Judging by how well today had gone so far, she didn't suspect their time at Ghostland would end so painlessly.

Colored lights flashed in the darkness in the dim light beyond the doorway. Ben lead the way into a large, high-ceilinged room filled with black metal cabinets, each housing a dozen racks of electronics. The flashing lights and LED screens illuminated hundreds of switches and knobs, wires and cooling fans. Cables snaked from the backs of them and gathered in bundles above the door, trailing back the way they had come: to the exhibits, to the control center, to every ghost and hologram and mechanism in Ghostland.

Each of them corrupted now, poisoned by Garrote's viral coding, his insane disease.

"What now?" Lilian asked. The hum had grown so loud she had to shout to be heard.

"Pull the plugs, if we can find them!" Ben shouted back.

They entered the maze of cabinets. A metallic clanging rose above the hum of electronics from somewhere deep within. It happened twice more before she heard an old man shout gruffly: "Christ in a sidecar!"

Ben turned with excitement in his eyes. "Someone's alive in here!"

They kept moving, further into the maze, the darkness alternately lit up green and red like Christmas. She followed the outside, her right hand lightly touching the smooth, cold surface of the cabinets. It was a method she'd often used in video games, aside from the touching. One she knew would inevitably lead them to the exit. After several turns, the clanging resumed, followed by another curse: "Son of a bastard!"

Much closer now. They had to be nearing the exit.

"You gotta be kidding me!" the man shouted.

This time Lilian was sure she recognized him.

"Stan? Is that you?" Ben shouted over the hum.

The detective's voice came back, filled with uncertainty. "Ben? You kids are alive?"

"We're alive!" Lilian called back. "Where are you? Keep talking!"

The detective stammered. "Uh, okay, umm... 'I am Javert,'" he sang tonelessly. "'Do not forget my name! Do not forget me, 24601—'"

Lilian followed Stan's voice until they reached an opening in the cabinets. This was not an exit. Judging by the high vaulted ceiling, which appeared to continue away from them for fifty feet or more, the large space appeared to be at the center of the maze.

Detective Beadle stood in the entrance gripping a fireman's axe. He was panting, sweating through the armpits of his shirt. His sport coat and fedora lay crumpled on the floor at his feet. Behind him the cabinets were laid out in an octagon, about ten feet in diameter. At its center a complicated-looking machine rose to touch the ceiling. Lilian couldn't see past the detective to get a good look at the base of it, but it glowed an unearthly shade of red.

"Whoa," the detective said. "What's with the get-up, Major Tom?"

"Are you really alive?" Ben asked, aiming the stun gun at him.

Stan shook his head in confusion. Then he broke out in laughter. "I'm sweating profusely, my aching joints are a constant reminder of my fragile mortality, and my piles itch like crazy. If this is what death is like, then by all means, zap me into oblivion."

Satisfied, Ben lowered the weapon. But Lilian remained cautious. "How did you get here?" she asked. "How did you find this place?"

"Good old-fashioned police work, that's how." He flashed a crooked grin, wiped his brow with a forearm, and explained, "When the shit hit the fan—pardon my French—I started asking anyone I could find wearing one of those red and purple Ghostland T-shirts. Turns out there's a secret exit down here that leads to a mechanical shed on the other side of the wall."

Lilian's heart leaped. She set her skepticism aside. "You know the door code?"

The detective smiled and tapped his temple. "You don't think I'd come all this way without it, do ya?"

Ben and Lilian shared a sheepish look.

"Anyways, I finally get here and lo and behold the door's wide open. And then I find *this*..."

The detective stepped aside, revealing a large glass cylinder within the machine.

And there he was, in the flesh.

Lilian couldn't tell if the thick, translucent liquid in the tank was red or if it had taken on the hue from the servers surrounding it. A fat, tight bundle of cables rose from a hole in the raised access floor and up over the top of the tank, coiling down into the red liquid, surgically attached to a naked, hairless man floating within: at the wrists, the legs and several directly into his shaved skull and torso, giving him the appearance of a modern-day techno-Frankenstein.

Even with his trademark mustache shaved off, Lilian recognized him.

It was Rex Garrote.

A slow grin appeared on the detective's wrinkled face. "Told you he was fakin' it."

"The water is red," Lilian said, in fear and awe. "Just like Allison said."

"How could she know about that?" Ben wondered aloud.

Lilian shook her head. The question troubled her, made her heart hurt as much as her brain. She said, "I don't know," and left it at that.

"You kids wanna fill me in?"

Lilian cautiously approached the tank, ignoring the detective. Garrote's body bobbed languidly in the thick liquid. Bubbles rose sluggishly from the bottom of the tank but not from his nose or his slightly parted lips. He didn't appear to be breathing. For all she knew he really was dead.

"Is he alive?" Ben asked.

"His brain is," Stan said. "Got him all wired-up live and direct to these computers, like some kind of human Duracell."

Lilian saw Ben nod thoughtfully in his reflection on the glass. "The Singularity," he said.

"The what now?" Stan asked.

"The Singularity. All those books in his library about humans merging with machines. He *did* it," Ben marveled. "He actually uploaded his consciousness to the computers, into the Ghostland system. That's why his hologram seemed more realistic than the rest of them. And that's why he was able to shut down the Recurrence Field and take over all the ghosts."

Stan blinked rapidly. "Excuse me?"

"The Ghostland people thought it was a virus," Ben explained. "But it wasn't a virus. It was Garrote, working his way through the system, gobbling up lines of code like Pac-Man and spitting out clones of himself. Lilian saw it first on the Ghost Tram. He was taking control of all the ghosts in the park, one by one, probably from the minute they turned on the system. He could see through their eyes, control their movements, make them kill."

Stan blinked again, seemingly unable to process what he'd heard. Then he turned to Lilian. "Is all that true?"

She nodded. "It's true."

"We have to unplug him," Ben said. "Before he figures out where we are. It's only a matter of—"

He stopped and stared, his lower lip hanging open. Lilian followed his gaze to the tank and jumped back in fear verging on panic, her skin crawling.

Garrote had opened his eyes.

Stan gripped the axe in both hands. "Oh, Christ, Frankenstein's awake. I've been battering on this glass for the past five minutes but the damn thing just won't crack."

"Have you tried the console?" Lilian asked.

"Console?"

She stepped up to the terminal, eyeing Garrote cautiously, certain he would reach out to spook her. One final jump scare. But he remained motionless, staring dead ahead, his glassy eyes blank. On the console was a digital screen with several complicated readouts that looked like they belonged on hospital equipment. One appeared to be brainwaves, another a heart rate—49 BPM—along with other numerical data and symbols Lilian couldn't decipher. On the upper right corner was the word **LOCKED**, surrounded by a thick black border.

Her hopes sank. "Even if we could figure out how this thing works, the terminal's locked."

Ben thought for a moment. "Let's just pull the cables," he said finally.

The detective shrugged and gripped the axe in both hands. He turned to Lilian, who nodded.

Garrote posed no objections, floating silently, staring into the middle distance.

"Do it," she said.

"*Wait!*"

The voice came from the darkened corridor, beyond the reach of the flashing lights, but Lilian recognized it immediately. Anywhere else, she would have sworn it was impossible.

Here in Ghostland, the dead came back.

Allison stepped out of the darkness, approaching them slowly. She bore none of the wounds she'd died of, and was dressed in a pale pantsuit, like the ones she'd worn in their sessions, its color impossible to determine under the flashing lights.

Ben crept back to the corner of the small space until he struck the cabinets. He startled, looking back over his shoulder as if he'd walked into a spider's web. Stan kept looking between the three of them, seemingly unable to decode the change in the social dynamic, why Ben and Lilian would suddenly be afraid of this woman who'd until very recently been their friend.

All the while a solitary thought repeated itself in Lilian's head, like a ghost caught in a Recurrence Field loop: *That's not Allison. That is* not *Allison...*

"Ah," the ghost said. "You're surprised to see me, aren't you? I would be too, if I were you. In fact, I'm a bit surprised myself."

Stan said, "I could use a little context, kids."

"She's dead," Ben told him. "She died in the asylum."

Stan looked momentarily afraid. But his features quickly returned to their normal expression of mild curiosity.

"I died," Allison said. "But I've come back to warn you not to open that tank. If you disconnect him—" Under the flashing red and green lights it almost seemed as if her face grew dark along with her expression. "—it's all over. For all of us."

Lilian felt herself on the verge of tears again, and bit her lip to prevent them. "How do we know we can trust you?" she asked.

"Lilian," Allison's ghost said, disappointment in her tone. "I'm so sorry I failed you—failed both of you. We were supposed to protect you, but we didn't. We *couldn't*. How could anyone prepare for something like this? The whole world turned upside down. Everything we knew, took for granted—all gone. But you have to know, after all we've been through, all the time we've spent talking—we have a sacred bond, you and I. I take that very seriously. Not even death could break it."

"Then tell me why I shouldn't pull the plug right now. Give me a reason." As Lilian spoke she gripped the bundle of cables at the back of the machine.

"Because that's exactly what he wants!" the ghost cried. "He wants to be *set free*, Lilian. As long as he's trapped in that body, he can't leave this place. If you unplug him, he won't need the Ghostland program any longer. He'll be able to leave this place and go anywhere he wants. Infect the whole world with his poisoned mind."

"But he needs us to open the door," Ben said.

"Why? You said it yourself, he's in full control of the park. Why would he need you if he could just open any door he wanted as easily as switching on a light?"

Ben turned to Lilian with a look of doubt. Lilian shared it. Could Allison be trusted? More important, was she *right*?

If they pulled the plug, would they be playing right into Garrote's hands?

The weight of the decision was too much to bear. She let go of the cables and stepped away from the machine, peering up at the motionless man in the tank, the man behind the curtain, wondering what the hell to do now.

LOOSE ENDS

W*HAT NOW?* BEN thought. He stood backed against the servers, his gaze flitting between Allison's ghost, his two living companions, and the living dead man in the tank.

Harrison had told them to unlink Garrote from the servers, but how could they be sure the programmer had been telling the truth? He could have been working with Garrote. He could have been *taken over* by Garrote, another ghost gobbled up by his code.

Whatever decision they made posed a huge risk. If they chose to stay here and possibly die to prevent Garrote and his army from escaping, the police would shut down the power eventually, break down the wall and potentially release them anyway. If they chose to escape, leaving Garrote linked to the computers, it wasn't certain his hold on the ghosts would be broken. If they killed Garrote right now, his code might simply break free of his body, as Allison seemed to suggest. Or it might die along with him.

What the hell do we do now?

Lilian interrupted his thoughts. "How did you know about the red water?"

The ghost seemed flummoxed by the question. "What?"

As she asked this, Ben saw Allison's face flicker. A digital glitch almost too quick to see, but long enough for him to make up his mind. For that split second, her face had become a dark reflection of the man in the tank. Garrote's mind virus had gotten to her, had made her his slave.

Whatever she said was not to be trusted.

"When you died," Lilian was saying, "you said Garrote wanted to drag us down into the dark, that he was waiting for us in the red water, and here he is, just like you said. How did you know?"

"I don't know how I knew, I just—" The ghost shook her head. "It came to me. Like a vision. I saw the water, blood-red just like that." She nodded toward the tank. A squadron of slow, thick bubbles rose from Garrote's nostrils and broke on the surface. "His eyes were closed at first, and then they opened and he reached for me. And right away I knew this was exactly where he wanted us. *In this room*. Because he wants you to pull the plug, Lilian! You have to believe me—"

"She's lying!" Ben said, pushing away from the wall.

As they turned to him, he crossed to where Allison stood, a ghost which no longer belonged to her former self, her image stolen like a dead celebrity dancing in a vacuum commercial, her thoughts and memories appropriated by the madman wearing her face, who used his control over her now to shake her head.

"I'm not lying," the madman said through Allison's lips, in Allison's voice.

"I saw Garrote," Ben said to Lilian, to Stan, pointing at the ghost. "It was quick, but I saw him in her face. He's scared. He wants us to give up. To run away. He wants us to open the door for him. Let him out of his cage."

Allison shook her head, weaker this time.

"But without the program, he's nothing," he continued. "Ones and zeroes in the shape of a man, just like Sara Jane Amblin said."

"You're wrong," Allison said softly, looking at the floor in front of her.

"Stan, cut the cables."

The detective seemed unsure. He hefted the axe in his hands as if testing its weight but he made no move toward the tank.

"If you don't cut the cables, I'll do it myself."

"Cut the cables, Stan," Lilian said, the hurt and confusion evident on her face.

"*Don't you do it!*" Allison roared, and reluctantly, not wanting to harm the woman within—if any part of her essence still existed—Ben reached out with the stun gun and squeezed his eyes shut as he pulled the trigger.

In the instant the charge hit her he was struck with the true bleakness of it. Allison had been the first to question the morality of this place, this carnival freak show. The first to object to using the stun gun against the ghosts, who were once living, breathing beings.

Now he was using it on her.

He tried to convince himself she was already gone but as she screamed, the shock surging through her, he couldn't help but feel guilty. Before he could take it back Allison's form exploded in a supernova of multicolored light, a swirling galaxy of particles. Then she was gone, and the three survivors—possibly the only living souls left in the entire park—stood alone with Garrote's body naked and prone inside the tank.

"Do it," Ben said.

Without a word, Stan hauled the axe over his shoulder and swung it at an angle, striking the cables trailing down the back of the tank. Sparks flew as the wires on the outside of the bundle severed cleanly. He swung until the blade cut through the last cables and struck the glass behind them. Their severed ends slipped into the tank as the thick glass finally splintered, cracks webbing outward from the point of impact. Garrote's mangled body struck the center and the tank shattered outward, the thick, goopy liquid pouring out in a red wave. Garrote spilled out with it, landing with a wet slap on the floor.

They stood around his corpse, wires like thick black snakes still attached to his limbs, his chest and head. They watched for breathing, for a twitch in his limbs, for him to spring up from the floor and attack.

None of this happened. His body remained still, his chest never rising, the slick liquid already beginning to dry on his goosebumped—*horripilated*—flesh. The lights on the servers continued flashing and the hum of their fans droned on. Only the terminal connected to the machine had changed: its screen flashed emergency red, the waveforms flatlined, the numerical readouts all zeroes.

After a long moment, Lilian finally broke the silence. "That's it? It's over?"

"You never heard the expression about looking gift horses in the mouth?" Stan said, letting the axe fall to the floor.

"No. It's not over yet," Ben said, standing over the writer's corpse. He found it hard to believe the man was really dead. The world had believed Garrote had killed himself longer than Ben had been alive. For them to discover the writer had been alive all that time, that he'd faked his own death, and then find him attached to machines, all within the span of hours—now that he appeared clinically dead Ben couldn't just let it go. He needed to be sure there was no life left in him, not even a trace of conscious thought caught somewhere within his lobes.

The only way to be sure was with fire.

He reached out with his foot and kicked Garrote's shoulder. The writer's body moved slightly, rising from the floor and falling back as Ben stepped away from him. Ben took a knee and checked the writer's neck for a pulse, the cold liquid slimy under his fingers.

"We should go," Lilian said.

Stan bent to pick up his sport coat from the floor.

"We have to destroy the servers," Ben said.

Lilian looked at him as though he'd lost his mind. "Why?"

"It's the only way to be sure." As he said it, he reached into his pocket and brought out the lighter fluid. "Wipe everything out."

Stan goggled his eyes at him. "Whoa whoa, hang on a minute. What are you doing, kid?"

"I'm gonna burn his body so he can't come back. Then I'm gonna smash the servers so no one can ever start this program up again."

"Burn him? Ben, let's let well enough alone—"

"He killed my friends, Stan! He tried to kill me, too." *And he almost succeeded*, his mind finished, dredging up a famous line from Richard Matheson's *Hell House.*

"I understand you're upset. But we need to get out of here—"

"We have to *go*, Ben," Lilian said.

"You go. I'll catch up with you."

"I'm not leaving here without you."

"Then stay. Help me end this."

He squirted a jet of lighter fluid onto Garrote's back, down his skeletal ribs and bony shoulders. Another on his bald head, over his ear and down his cheek. He squeezed the bottle methodically, again and again, a blanket of Zen calm draping over him. He thought he'd never felt so calm in the past four years, not since the day he'd seen Rex Garrote standing in the window, watching him.

"Come on, kid," Stan pleaded.

Ben ignored him, splashing the accelerant on Garrote's weak, atrophied arms, across his wrinkled buttocks and spindly legs. The smell of the lighter fluid prickled his nostrils. His limbs felt light and airy. It wouldn't have surprised Ben to know he was getting high, whether from the fumes or the knowledge that Garrote would soon be gone, dead, *finito. Dig it, babies,* he thought, in his best Rex Garrote impression.

"Ah, fuck it," Stan said.

In his peripheral vision, Ben saw Stan bend to pick up the axe, heard its heavy blade scrape on the floor tiles. The detective brought it to the servers, hauled back, and swung.

The blade struck one of the machines and the plastic splintered, parts scattering in every direction, exposing diodes and capacitors and flashing LEDs. He swung again and sparks arced out in a shower of plastic and machinery. The room dimmed slightly as its lights winked out.

Ben squirted the last of the can onto Garrote's body and tossed the tin aside—remembering the madman performing a similar action with the gas can in the library. He reached into his pocket for the Kitchen God matches and froze, his whole body seizing in horror.

The pocket was empty.

"No," he muttered, his plan ruined in an instant of carelessness. He'd lost them. Somewhere along the way they must have fallen out of his pocket. "Where are they? *Where are they?*"

Lilian said, "What happened?"

Behind him Stan had gone to work on another server. He heard the whoosh of the blade slicing the air and the car crash jangle of plastic and metal.

"I lost the matches!" he cried, getting angrier with himself, more frustrated. "I lost the fucking matches!"

A hand fell on his shoulder. Stan had laid the axe against the servers and was reaching into the lapel pocket of his jacket. "Loosen up, kid," he said. He took out a chrome Zippo lighter. Ben snatched it eagerly from his hand. "I don't need it back. Burn that son of a bitch, Ben. Send him straight to Hell."

Smiling, Ben spun the wheel. The flint sparked but didn't light. It caught on the second try, flickering with an orange flame that warmed his hand. He glanced over his shoulder at Stan, and the retired detective gave him a nod of encouragement. Lilian watched him with a mixture of fascination and dread. He supposed, with the firelight flickering over the shadows on his face, he looked a bit like a madman himself.

"You might want to step back," he told her.

She took two deliberate steps backward, never taking her eyes off of him.

Ben tossed the Zippo at the dead man and stepped back in a somewhat clumsy movement. The flames engulfed Garrote's body—for a moment, Ben was sure they had all been fooled, that he'd been squirting lighter fluid at a hologram and Garrote, the *real* Garrote, would step out of the shadows to applaud their futile attempt before closing the curtain on them for good.

But Garrote lay there, his flesh turning a crisp brown, bubbling and crackling with blue-tinted flames. Ben stared at the macabre sight, mesmerized, amazed he'd come so far, that he'd lived long enough to see this insane plan through, to burn his former idol to ash and watch the man's dream crumble to dust.

Stan struck another server with the axe and this time it felt as though the ground beneath their feet had shaken. The sensation pulled him out of his reverie and he realized with sudden terror that his heart was beating very rapidly. He reached into the hip pocket of his shorts and felt for the few pills he'd managed to save from the poltergeists in the prison. He popped two into his mouth, worked up some saliva and swallowed the bitter pills whole.

The ground shook again. This time Stan held the axe over his shoulder, about to swing. The server cabinets rattled. Stan looked back as sparks fizzled from the half dozen machines he'd damaged, their electronic guts spewing from their shells.

"Did you feel that?" Lilian asked, her eyes wet and wide with fear, covering her nose from the smell.

Ben looked at the burning man on the floor, a mass of charred flesh, split open and cracked, almost unrecognizably human. Fat and blood oozed out, bubbling and blackening to a thick scum on the tiles. The smell was both terrible and savory, something like burnt pork, with a cloying, rich undersmell of copper, scalded urine and sulfur.

He nodded. "Let's go."

Stan kept the axe and the three of them stepped into darkened maze of servers. The ground shook again, hard enough to loosen plaster dust from the ceiling and make several cabinets swing open in their path. The survivors danced delicately between them.

Before they rounded the first corner, Ben risked a glance back at the room they'd just left. Fire flickered over the servers, and shadows moved in the yellow light. But Garrote was dead and ghosts threw no shadows. He passed it off as a trick of the light and kept following behind Lilian, who followed behind Stan, still holding the axe out in front of himself as if he thought it would protect them from ghosts.

The floor beneath their feet rumbled as something high above them crashed to the ground. Ben looked up. In the alternating green and red light, he thought he saw cracks in the ceiling. Whatever had fallen had been immensely heavy. Was Garrote House itself coming down? Ben hoped that it was—though the timing could be better. He wished he could have seen it for himself, even from the other side of the wall. Down here it meant nothing to them but more danger.

"I'd really feel better if I led the way," Lilian said.

"Nah," Stan said. "I'm closest to death, I should be up front."

"Then at least follow my directions."

Stan stopped at a T-junction and turned to Lilian. "All right then. Which way—?"

He gasped. His whole body shuddered. His eyes went wide and the hat canted then fell off his head. His fingers unfurled from the axe handle and the blade clanged to the floor, the haft toppling.

Ben thought, *Heart attack?*

Lilian called Stan's name. He didn't answer, merely gripped his stomach. Pain, sorrow and surprise suddenly fighting for control over his facial features. A mouthful of blood spurted from his lips and dribbled down his chin. More blood oozed out from between his fingers and down over his shiny belt buckle.

Ben grabbed Lilian's shoulders and pulled her back as the detective slumped against the cabinets and slid to the floor with a thump, his blood streaking the black metal, his eyes glazed, looking up at his killer.

The man stepped over him, a long knife held in his right hand, dripping with the detective's blood, his blue chambray shirt and khakis stained by gouts of it. Handcuffs swung from a scabbed and bruised wrist. He smiled.

"Told you I'd stick you, pig," the Doll's Head Murderer grunted, tearing off his headset to get a good look at his victim. He worked up a mouthful of saliva and spat on the bleeding detective. The thick gob spattered Stan Beadle's eyes. He sat there, unblinking, as the murderer's spit oozed down over his eyelashes and onto the lines of his stubbled cheek.

Then the light faded from his eyes for good.

Alex Fischer, still very much alive, turned his handsome, youthful grin on

Lilian. "Hey there, dollface," he said with surprised delight. "What's shakin'?"

Above them the ceiling rumbled again, louder and longer than before. And in flashes of intermittent dark, the man with the knife began to laugh.

THE HOUSE FALLS

W*E'RE DEAD*, SHE thought. *We're so close, almost there, now it's all over, he killed Stan, he fucking killed Stan!—no code, no way to escape, no way to stop him, nowhere to run, and how did he get here, how did he get here, how the hell did he FIND US?*

"I'm so happy to have found you, Pretty Polly," the murderer said, swishing the blade back and forth like a conductor's baton. He stepped over the detective's body, coming toward them, and Lilian back-stepped, bumping into Ben's shoulder.

"Don't come any closer!" Ben shouted. In the confined space beneath Garrote House his voice thundered.

The killer grinned. "Or what? You're gonna zap me?" The grin widened. "Well, go ahead. Try it. I'll even give you the advantage." He slipped the knife from his right hand to his handcuffed left and tucked the right behind his back. "You feeling lucky, kid? 'Cause I am. You better believe I am." He jiggled the handcuffs hanging from his wrist. "If you idiots hadn't lured that dumb gangster over, I never would've got away from the bitch's trailer. And that four-eyed doufus, that nerd in the computer, he led me right to the pig with promises of sweet revenge."

Four-eyed doufus, Lilian thought, sharing a brief look of despair with Ben. Could he mean Harrison? And if so, had they really made the right choice severing Garrote's connection to the computers? Had they killed him —or just set him free?

The killer was still blathering. "And then ho-ho-ho! Old Man Stan walked right into my web with two little stinkbugs in tow: the turd and the tartlet.

One to kill and one to..." He winked at her. "Well, that's for me to know and you to find out, sweet cheeks."

"Shut up!" Ben said, thrusting the stun gun forward. "Don't talk to her like that."

Fischer gave him an impressed look. "Well, we've just witnessed a historic moment here, sports fans!" He turned the knife on Lilian. "Looks like your boyfriend's balls just dropped."

Sonic booms shook the tunnel before she could reply, one after the other, rattling the cabinets on either side of them and sending down another shower of dust, this time speckled with bits of concrete that cracked and crumbled on the floor around them.

They had to get out of here, and quick. The crack in the ceiling was becoming a chasm. The next crash could topple the servers and collapse the tunnel and they would all die down here in the dark—or worse, they'd survive, broken and unable to move beneath the rubble of Garrote House, until the ghosts found them or they died of starvation.

"Boy oh boy, it's all coming down up there," the killer said with very little concern.

"We have to get out of here!" Lilian shouted at him.

The killer shrugged. "Maybe. But we're gonna have us a little fun first. Yes, we are."

He advanced another step and suddenly the solution came bright and clear in Lilian's mind. But it depended on Ben being quick with the stun gun and he had a dazed look in his eyes, which likely meant he'd taken another pill. If his response was delayed, the stun gun wouldn't matter. Fischer would gut them both and—if the programmer really had led him here—walk right out the back door, letting Hell out with him.

There was no other choice. She had to try.

Fischer took another step. One more and she would spring her trap.

She hurled a thought at Ben, hoping the intensity of her emotions would somehow push it from her mind to his with very little belief that it would: *Follow my lead*, she thought, watching the killer's shoes. It was a dance between three people. *One and two and...*

Fischer stepped forward.

Lilian grabbed the cabinet door and swung it outward.

The blade struck it, scraping along its metal face with a high-pitched squeal. The killer's hand jerked backward and the door struck him in the face. He screamed, the sound muffled as he grabbed his nose. Catching her cue, Ben shot out with the stun gun, pulling the trigger, angry blue-white sparks jumping between the electrodes—

And Fischer batted his arm away easily. He kicked Ben in the shin with the heel of his brown leather Oxford and Ben dropped the stun gun, falling to a knee.

"Oh, you little bitch!" Fischer said, his voice nasal, his nose mashed and oozing blood into his mouth. He slammed the door closed and kicked the stun gun out of reach. It skittered across the floor and stopped between Stan's legs. "Oh, you are gonna *get* it now—"

A terrible crack came from above, as loud as a shifting tectonic plate. Lilian looked up but Fischer ignored it, he kept coming toward her with the knife, even as a fresh rain of concrete and dust sifted down, blood staining his teeth and streaming down his chin like he'd just eaten a meal of raw flesh, his lips curled upward in a satisfied smile.

The chunk of ceiling struck Fischer dead on, and in the instant before the floor collapsed beneath his feet Lilian saw his head flatten, wiping away his handsome rich boy's smile with a remarkable splatter of brain and skull and teeth, and his bones—every one of them—pulverized and flattened as easily as if the clothes he wore had been full of stuffing, and all that was left to prove he'd just been standing there and that the entire ordeal hadn't been a hallucination or simply another hologram was the single Oxford shoe sticking out from the shallow pit beneath the giant hunk of concrete and bent rebar.

"*Holy shit*," Ben said, looking down at the rubble barely five feet from where he stood. He looked up at the black hole in the ceiling and laughed anxiously. "Holy shit, did you see that?"

Lilian said, "Come on."

They stepped around the debris. Ben stooped to pick up the stun gun at the detective's feet. He snagged Stan's hat with his other hand and laid it on the old man's head.

No time for a moment of silence. Another boom tipped the cabinets behind them and they fell like dominoes, crashing down over the killer's remains. Lilian ran down the corridor to the right, hoping that if they were forced to backtrack, they would still have time before the whole damned ceiling collapsed on them, covering them in the rubble of Garrote House.

Another boom resounded in the semi-dark. The brittle *CRACK!* that followed, like lightning preceding the thunder, started somewhere behind them and radiated outward in multiple directions, splitting the concrete along jagged fault lines. Server cabinets toppled in their wake—*BOOM! BOOM! BOOM!*—as Lilian and Ben leaped out of the maze into a large open space the size of an empty parking lot, its circular pillars cracked and straining against the shifting weight above.

As they looked back, a portion of ceiling roughly the size of Kansas caved in, smashing heavily on the server maze with a sizzle of electricity and a mushroom cloud of dust and debris.

Ahead of them, a seemingly impossible distance away, stood a small alcove, its concrete wall painted with diagonal yellow safety stripes, a cone of white light shining down on the large metal door.

The hatch.

It was real. They'd found it.

Behind them the antique carpets, flooring and furniture of Garrote House tumbled noisily into the chasm, crashing down on the mountain of broken concrete and jagged metal, tables and chairs splintering, china and stemware smashing, a grand piano obliterated in a discordant explosion of black and white ivory and a twang of snapped strings. The chandelier jingled the whole way down and struck the expanding ruins with a dull, unmusical thud, followed immediately by an entire flight of stairs, which collapsed accordion-like into itself, then the fireplace hearth and a portion of chimney, still laden with burning logs.

The entire process happened quickly, and the walls came down in *Tetris*-shaped chunks of brick, letting the last dying light of day shine through the ragged opening.

"Ha! Yes!" Smiling wide, Ben leapt into the air, pumping a fist. When he landed, he grabbed his chest, bunching up his T-shirt between clenching fingers. "We did it," he said, his brow knotted, his voice almost a groan. "We beat him, Lilian! We fucking *beat* him!" He was so thrilled, so over the moon, he barely seemed to notice he was in pain.

Silence. The great collapse had subsided. A few small cracks sounded intermittently here and there, like lake ice expanding, but it seemed the worst was over.

But she knew they weren't safe yet, and as she thought this, almost as if her mind had conjured them up from nothing, the first wave of Garrote's army began streaming through the massive hole.

THE HATCH

WITH A FIST pressed between his ribs to keep his heart from bursting out of his chest, Ben stood marveling at the ghosts pouring through the jagged Hellmouth into the tunnels below.

For a moment he wasn't sure it was just ghosts in the exodus: among the legions of dead prisoners and soldiers and carnival workers and pirates and civilians from every other period in history, there were too many people wearing modern attire, survivors in flip-flops and fanny packs and fullback ballcaps and T-shirts with popular catchphrases and iconic fictional characters. These were Ghostland customers, the recently deceased. Ben watched them drop through the hole like paratroopers into enemy territory, sunlight filtering through their bodies rather than catching them in silhouette, and a creeping sensation crawled beneath the flesh of his neck and cheeks.

Scant seconds ago, he'd been sure it was all over. Now he could see it would never end. By separating Garrote from the machines, they'd allowed him to escape the confines of the program. None of the technology his people had made was necessary any longer. In death, for-real death, Rex Garrote had evolved beyond the program.

Beyond Ghostland.

He'd won.

Lilian grabbed Ben by the shoulder, trying to pull him along. But he stayed put, watching the surge, a feeling gnawing at the edges of conscious thought like a caged animal, a sense that he'd forgotten something terribly important.

He could see Lilian was starting to panic, caught between bolting for the door and waiting for him to snap out of it. And somehow, aside from the hammering of his heart, Ben felt calm, almost at peace. All he could think was that the fumes from the lighter fluid must have really gotten to him. He felt lightheaded, and the strange, floating sensation was traveling from his brain down his arms.

"Let's get out of here!" Lilian shouted above the noise.

As the wave of dead energies flooded down the mountain of rubble, over the last remnants of Garrote House—making Ben think of ants, of lava, their chanting and cheering and howling and yammering melding together into an ocean of sound—Ben chased Lilian across the divide between the coming horde and the possibility of escape, of freedom beyond the hatch, beyond the wall. His limbs felt loose and rubbery, like someone else was piloting them. Numb. He was tired. All he wanted now was to get out of here, to go home, and take a long, well-deserved nap.

They reached the alcove surrounding the hatch. Still running full-tilt, their palms slapped against the wall, slowing their momentum, preventing them from crashing headlong into it. Hot white light shone down on them. They were completely out in the open now, completely vulnerable.

He turned to look. The dead had filled the—*Arena*, he thought—concrete space from side to side. An endless sea of faces, shimmering in the dim light. And suddenly the wave stopped surging forward. The ghosts hovered motionless, maybe fifty feet from where Ben and Lilian stood under the light above the door.

"Now what?" Lilian said, shouting to be heard over the noise. The door looked like something that belonged on a spaceship, with rounded edges, a tight seal around it, and a wheel-lock in the middle. She was looking at the keypad beside it, a digital readout with keys lighted green, consternation crinkling her brow. The code wasn't numerical. The tiny black print seemed to be from some symbol-based language, but Ben couldn't decipher any of it: it could have been Chinese characters as easily as ancient Sumerian. As he looked at the images they all blurred together. He felt the crushing weight of desperation in his heart.

Lilian thrust her fists against the door, crying out for help. Ben wished he could have saved them. But it was too late. The metal sounded incredibly thick. Likely soundproof. Anyone on the other side might hear a dull thud, a muffled sound something like a cry. Without a headset, they wouldn't hear the stadium-sized roar of Garrote's army.

All was lost.

He turned to face the ghosts, embracing his impending death. In this place of sudden peace, he spotted familiar faces among the surging crowd. There were Niko and Leonard, side by side at the front of the charge, their eyes as blank and emotionless as Morton Welles's and the nun's, several feet deeper into the throng. There was the dog, Freddie, and Harrison the programmer—who hadn't survived, had clearly just been used by Garrote to get what he wanted—there were Sara Jane Amblin and Dr. Death, there was the Ice Cream Man, who had finally recovered his white paper hat, and Detective Stan Beadle very near to him, and a cowboy studded with arrows, and the boy from the ticket booth. The Japanese teenagers all huddled together. The cigar-chomping gangster. The doll on the tricycle. The Behemoth. Ben saw them all, like an instant replay of his last day on earth.

Darkness fell over the ghosts as some massive thing rose above the pit in the earth. First came swirls and tangles of dark brown tinged with gray, followed by a peach-orange landscape of ridges and small black craters. As two scrubby chestnut bushes appeared over the chasm, the odd-colored terrain furrowed between them, Ben suddenly realized what it was—and all of the fear he'd ever experienced in his entire life paled in comparison to what he felt now.

Rex Garrote towered over them, the Shōki himself, the last living ghost peering over the great heap of rubble that used to be his home. He was at least one-hundred feet tall and completely free of the limitations of his body —*All thanks to me*, Ben thought miserably. He was a sadistic general huddled over his troops, a demented puppet master, a titan waging war with human toys. Worse, he was pissed.

"Oh, that's not good," Lilian said.

The writer grinned, each of his teeth as tall as his ghosts.

In his terror Ben experienced a rare moment of complete clarity, the kind of epiphany that comes only a handful of times in a lifetime, if at all. He remembered suddenly what had been troubling him and knew exactly what they had to do next.

"Lilian," he said.

"What?" She was distracted, still pounding on the door, weaker with every strike. She paused briefly to look at him, then continued the futile gesture.

"If they've come down here," he said, "that means they can't get through the wall. Garrote can't shut down the Recurrence Field on his own. He still needs us to open the door!"

"What are you saying?"

"I'm saying we have to give up, Lilian. It's the only way to stop him. If we let him out, he won't stop until everyone is dead. Our families. Your friends. Everyone in Duck Falls. In America. The *whole world*."

Anguish made her face quiver. "Ben..."

"I know." He gripped her shoulder. "I know how it sounds. But it's the only way—"

"You don't know that. He could lose his power out there."

"I wish we could be sure of that. But we can't take that risk, Lilian. We just can't."

She looked out at Garrote's army, at the monstrous man who'd killed so many in such a short span of time, who'd infected them all with his mind, with strings of code, ones and zeroes. They would kill for him. They would kill and kill until nothing good was left in this world.

"He'll kill both of us, Ben."

"I dunno," Ben said. He was grinning, but the fear in his eyes was clear. "We just have to survive long enough to trigger the next cutscene."

"The next cutscene," she said.

She remembered the words that had flashed across the screen all those years ago when Ben died: *Second Player Has Disconnected*. Was that all death was? A flash of pain and then *snap*—the system shuts down? It was difficult to think of all the death she'd seen today as a switch flicking from On to Off. But looking at it this way almost seemed to make it easier.

Because the switch turned both ways, didn't it?

What was Off could be On again.

If nothing else, all of this ridiculously expensive carnival-ride technology proved that.

As Lilian's eyes narrowed to slits, Ben thought he saw the shift in them—from desperate defeat to fierce determination—and he felt incredibly proud of her bravery. The choice was easy for him. He was likely to die here anyway. But Lilian had a bright future ahead of her. She had survived so many horrors today and now he was asking her to let go, to cast her instinct for survival aside on a hunch.

"Okay," she said. "You know, my parents always told me I'd do big things with my life. I bet they never thought I'd save the world."

Ben grinned again, a little less anxiously. "When people tell the story about what happened here," he said, "I hope they get it right."

Lilian stripped the keeper glove off her right hand and held it out to him. Somehow, he managed to raise his hand despite the heavy blanket of numbness weighting him down, and he clasped his fingers around hers.

"You're the best friend I ever had, Ben. I really do love you."

"I love you too, Lilian."

She smiled through standing tears, a smile both remarkably sad and infinitely happy. "You can call me Lil," she said.

He smiled back, squeezing her hand tighter, considering how fortunate he'd been to have shared even a small piece of his life with her. They pulled in closer together, shoulder to shoulder, blocking the door. They were partners. Partners forever.

And partners never gave up on each other.

With a flick of his enormous wrist, Rex Garrote sent the first wave of ghosts hurtling toward them.

EPILOGUE

PLEASE PLAY AGAIN

Six months later.

LILIAN PRACTICALLY CARRIED him through the first few levels of *Infinite Zombie 4K.*

Blake wasn't anywhere near as good a gaming partner as Ben but he took instructions well and didn't whine like Ben sometimes did while getting his ass fed to him by a buttload of hungry zombies. Lilian thought Blake was cute and he had the approval of all her new friends at Stanford, a definite bonus.

They'd met at a team-building event during Frosh Week—in a locked-room mystery, of all things—and had clicked almost immediately. They both came from small towns, both enjoyed video games and horror movies and both had a savage, sarcastic sense of humor. Early in their relationship, Blake had told her he would have been at Ghostland on opening day if his friend's car hadn't broken down that morning. When she'd gone quiet on the subject, he hadn't pressed her. He'd been patient and understanding. He'd told her if she ever wanted to open up about that day, he would be there for her. And if she *never* wanted to talk about it that was okay too.

It had only been six months since what the news had called "the Ghostland Disaster" but much of the day's events had already begun to feel like the fading memories of a bad dream. Over two-thousand people had lost their lives that day, making it the single biggest tragedy on American soil since 9/11. Lilian had been one of only one-hundred and eight survivors. If the hatch hadn't opened when it had—an event even the police who'd been working on the door with an acetylene torch still couldn't explain, passing off as a "miracle"—she often wondered if anyone would have made it out there alive. If she and Ben hadn't held off the onslaught in those last moments, there might have been no one left to tell the tale—or to tell the tale *to*.

For a few weeks all everyone wanted to talk to her about was what had happened on Ghostland's opening day. *Should be dead*, they said of her. *Sees dead people*. By the end of summer, after the funerals and candlelight vigils and having to refuse dozens of interview requests—claiming the events had been a blur, that in the trauma of the day, of losing people she cared about, she remembered very little—eventually people had stopped asking. There were others willing, often *eager*, to tell their story of survival, including Miss Delyse, a former TV psychic who had managed to hide in the tram car ride with her eight-year-old granddaughter until help had arrived. And the Ghost Brothers were more than happy for the extra attention. There were rumors they might be planning a "Return to Ghostland" special for the new year.

In truth, Lilian remembered everything.

Some nights she still woke in a cold sweat, feeling cold, dead hands crawling all over her, the hair on her arms and the back of her neck standing as if from the static electricity of the keeper suit. Other nights she lay awake in the near-dark of her dorm room trying to conjure the faces of friends she'd made and lost that day.

Ben had come to her during one of these nights, while the tinny, bass-less beat of her dormmate Abigail's music trickled out from oversized headphones and Lilian's alarm clock flashed 12:00 as if the power had just gone out.

When she'd opened her eyes, Ben had been standing at the foot of her bed, as though her dream had summoned him. His form had wavered slightly—it must have been difficult for him to hold, like the weak signal from the old rabbit ears they used to have on the basement TV—and she'd been able to see the clock through his waist. But he'd been unmistakably *there*, wearing the same clothes he'd died in, even after she'd blinked and rubbed her eyes to be sure she wasn't seeing things.

She'd sat bolt upright and flicked on the bedside light. Abbie had groaned and rolled over to face her, squinting at the light, asking her *what the hell, Lilian*. Abigail couldn't see Ben. That much was obvious. But to Lilian he was as real as anything. And when he spoke, she began to sob quietly.

"I told you I'd haunt your ass," he'd said with a smirk.

It had taken all of her strength not to leap up from her bed and hug him right there in front of Abigail, her arms encircling thin air in the shape of a boy who would never grow up to be a man.

"Hey, don't cry," he'd told her. "Not for me. I can go anywhere I want to now. Do anything. I'm freer now than I've ever been."

She'd broken down then and he'd laid a comforting hand on her shoulder. She swore she could feel it. He'd told her all about his adventures, traveling the world meeting others like himself. When she'd asked him what he meant by *like himself* he'd said, "Ghosts," and grinned. "You can say it. I won't be offended."

Blake paused the game and got up from the mattress. He kissed her forehead, cupping the back of her head the way that had always given her goosebumps. "Gotta go, boo. Early class tomorrow."

"Oh sure, just leave me hanging," she said with a grin.

Blake smiled his half-awkward half-handsome smile and backed out of the room, nearly bumping into a pack of girls giggling their way down the hall. He pretended to wipe sweat off his forehead at the near-miss.

"I like him," Ben said once Blake had disappeared around the corner. "Me and the other ghosts don't think he should say 'boo,' though. That's our word."

"I'm working on it," Lilian said with a smirk. "Wanna finish off this level?"

"Sure thing, partner." Ben picked up the controller and sat down beside her at the foot of the bed. The bedclothes didn't move, the mattress remained full. Anyone who happened to walk by would have seen Lilian sitting on the edge of her bed with the second controller levitating two feet to her right. But Ben was careful. He seemed to have a sixth sense about these things, and she had only ever been caught talking to him—talking to *herself*, to an outsider—twice.

Ben picked up the game right where Blake had left off, the two of them cutting down zombies with the meager weapons they had at their disposal.

"I don't want to scare you," he said after a while, mashing buttons on the controller, "but he got out."

"He...?"

"*You know who.*"

The swarm of zombies overtook her as she turned to face him, the crawling fear from that day returning, clutching coldly at her heart. "What are we going to do?" she asked, though she was afraid to hear the answer. She couldn't bear to think about him. She thought if she had to face him again, she would surely die.

"I have to disappear for a while," Ben told her. "When I come back, you need to be ready. We're gonna need your help."

"We?"

"The other ghosts and me. What happened at Ghostland wasn't just a disaster." His eyes grew very large—with fear or excitement or both, Lilian couldn't tell. "It was a *first strike*. Garrote wants a war, Lilian. A war between the bre—" He'd almost said *breathers*. She knew what his kind called people like her. Many of them held contempt for the living, the breathers. Ben had been trying to change that. "—between the living and the dead," he finished. "I can't let that happen."

"No," she said, feeling terror pulsing through her veins. But also, something else. She thought it might be a need for vengeance. Somebody had to pay for what happened. The Hedgewood Foundation—Ghostland's parent company—was tangled up in multiple class-action lawsuits and criminal negligence charges, but that had never felt like enough to Lilian nor—according to the news—the families of the deceased.

Someone had to pay. And as terrified as the prospect of being pulled back into the fray made her, the alternative frightened her more.

"You can count on me," she said.

"Who are you talking to?"

The voice came from the hall, causing Ben to vanish and his controller to drop onto the bedspread with a light thump.

Lilian turned to look at the girl sneering at her from the hall. She recognized her from the snowboard team and considered a lie—but what would be the point? Soon it wouldn't matter, not if Ben was right and Garrote's first strike became a war.

"I'm talking to a ghost," she said, challenging the popular girl with a hard look. "You got a problem with that?"

The girl's upper lip nearly touched her nostrils. She muttered *freak* under her breath and wandered off.

"Yeah, I am a freak!" Lilian called after her. "A freak who's gonna save your life despite her better judgement," she muttered to herself, picking up the controller Ben had dropped and placing it beside the console.

She allowed herself a tiny smile of satisfaction, thinking about how much she'd changed since the day Ben had chased her through the alley after school, only wanting to be her friend again, to get his partner back. How his lust for life had helped her take back her own, and how in death, Ben Laramie had lived more than he ever could have when he was alive.

Her smile faltered.

He got out.

Cold terror gripped her, made her shiver.

Rex Garrote was out there, amassing an army large enough to take on the entire world.

But now there were others who would stand against him. And when the time came, Lilian would be among them.

She picked up her controller. The fear had subsided. Ready to play again —this time in Hard Mode—Lilian hunkered down in front of the game and pressed ***START***.

PROLOGUE

GAME OVER, he thought, blinking through the cracked lenses of his glasses. The entire park was in full meltdown, and the brats had just left him behind, escaping with the psychiatrist. Now he was broken and probably dying, trapped in the control room with that insane, frenzied *thing* they'd called *Alpha* in the lab, but the kids had called *the Swarm*. It had already murdered Ms. Amblin, pinning her against the monitor wall like a heretic nailed to a cross. Any second now it would be coming for him.

"*HARRISON*."

The voice boomed over the blare of the alarm and the sound of chaos and destruction, driving a spike of terror into Harrison's heart. It was *his* voice. He was here in the control room, among the dead and the dying. He was *here*. Was it even possible, after everything that had just happened? How could *anyone* have survived?

Of course, it's possible, he thought. *All those long nights I spent coding made it possible. This... all of this... it's as much my fault as it is His*.

"*HARR-I-SONNNNNN*," the voice said again. *His* voice.

Harrison Greely raised his head, the pain in his neck and lower back and broken nose causing stars to shoot across his vision. The emergency lighting flashed in the cracked, dandruff-flecked lenses of his glasses. He took them off and wiped them on his shirt, streaking them with blood that had spilled from his nostrils. He settled the glasses back on the bridge of his nose and blinked into the erratic light.

Hiding somewhere in the front row, Nia Dearwood—with her dull pink hair and fake-vintage Pokémon T-shirt—let out a prolonged, high-pitched scream. Was she still at her terminal? *God, it'd be just like her to die at her computer,* Harrison thought. Not that he was any different, he supposed.

He raised himself onto an elbow, feeling weak, every muscle like wet spaghetti. All he wanted was to lie back down and sleep for a week, even if it cost him his life.

"*GET UP, HARRISON.*" Mr. Garrote's voice boomed over the comm system speakers. "*GET. UP. YOUR WORK HERE ISN'T FINISHED JUST YET.*"

The programmer rolled over onto his butt, managing to prop himself up on shaky arms.

"*There you are,*" the writer said, lowering his voice. "*Welcome back to the land of the not-quite-living.*"

The control room looked like an explosion in an office supply store: a mess of broken computers, loose paper and pens, a shoe with a foot still in it —*God, that's Bishop's Reebok, isn't it?*—torn off above the ankle, keyboards and ergonomic LED mouses, desk toys and orthopedic office chairs.

Sara Jane Amblin herself lay sprawled on the floor in front of the monitors. The entity that had murdered her—*Alpha, the Swarm*—still hovered over her remains, twisting itself back and forth as if examining her, looking for signs of further life to drain.

Above them, Garrote's massive face was displayed in mosaic over the entire wall of monitors, missing only the screens blacked out and cracked by the impact of Ms. Amblin's body. He looked like a religious maniac who'd painted his face with a giant black crucifix from forehead to lips, the paint chipped and cracked. His dark grin filled the second row from the bottom.

"*Hello again, Harry. Did you miss me?*"

The writer bellowed laughter. Everything was a big fat joke to Mr. Garrote. He was probably having a blast, laughing it up in that fish tank of his in the labyrinth of servers beneath his creepy old house.

"You crazy boomer *asshole*!" Nia cried, peering up at Garrote from the front row of computer terminals, accusation in her eyes.

Alpha Entity darted toward Nia at the sound of her voice, writhing and pulsating, leaving Ms. Amblin's remains behind. Nia leaped to her feet and ran, tripped over a desk chair and sprawled across the aisle. Panting, she rolled over and screamed one final time as the Swarm—as *Alpha*—descended upon her.

Harrison closed his eyes. He didn't like Nia. He didn't like any of his former coworkers, but having seen the Alpha suck the life out of Ms. Amblin made him realize he didn't dislike any of these people enough to delight in watching them die.

Her scream ended abruptly, almost as if Garrote had pressed Stop on a Halloween horror sounds tape.

"*That's better now, isn't it?*" the man himself said, his dark grin widening into a smile. "*We can hear ourselves think.*"

Harrison opened his eyes again, cautiously. The Swarm hovered between himself and Nia's dead body, convulsing like an unstable element, a dark mass of dead energy hungry for the life pumping through his veins.

For a moment he wondered if the featureless faces swirling within the cloud were scrutinizing him, sizing him up, the way they seemed to have studied Ms. Amblin's corpse only moments ago. He had no doubt Garrote would let it do the same to him, if he didn't follow the man's orders to the letter.

"I did what you asked! You promised you'd let me live!"

"*And so I shall,*" the face spread across the monitor wall said. "*But we have much to do before nightfall, Harry. There's still the little problem of getting me out of here and that, unfortunately, will require some assistance. The children who just left—can you keep track of them?*"

Harrison nodded. "As long as their headsets are on, they should be easy to ping. If not, we'll have to use facial recognition from the security cameras—" He frowned. "You know, this would be a lot simpler if you didn't *kill everyone* who could have actually helped me!"

"I'm *helping you, Harry. The rest of the team were an impedance to our Great Work. We'd have to explain how we got from there to here, and you know I find backstory oh so tedious.*"

Harrison bit his tongue.

"*Find the children, Harry. I want you to keep an eye on them. Don't do them any favors, though. When they survive, I want it to be by the skin of their teeth. It can't be made easy for them. Life isn't easy. Nor should death be.*"

"Okay," Harrison said. "But why them? Why not someone stronger? More capable?"

The smile stretched across the screens grew wistful. "*Because the boy was my Number-One Fan, once. I believe he'll do what needs to be done without even being aware I've been pulling his strings. And the girl has moxie in spades. She reminds me of myself, in my youth.*"

Whatever "moxie" was Harrison didn't think he'd seen it. She'd seemed like a typical stuck-up teenaged girl to him. A spoiled princess.

"But I could do it so much easier," he said. "I know the tunnel system, the server grid, the hatch code.... I could be with you in less than an hour, depending on—"

"*Tut, tut, Harry, my boy. You're far too delicate. My park would eat you alive. You'll be safe in here, as promised. And I need you at your terminal. I'll have my hands full with all of my new toys.*"

That smile again, as wide and deadly as a California fault line.

"*Oh, there's a man here today by the name of Alex Fischer. The twisted little freak is lingering around the Transportation building, by the mobile home exhibit. I suspected he wouldn't be able to stay away and indeed, I was right. I want him, Harry. I'll send his image to your terminal.*"

"What's so special about this guy?" Harrison asked, burning with jealousy. First the kids and now this Fischer guy. It was almost like Mr. Garrote didn't think of him as a partner at all.

"*Nothing much,*" the writer said offhandedly. "*I simply despise loose ends. Get busy now, my diligent little beaver.*"

Harrison pushed himself shakily to his feet.

As if attracted by his movement, the Alpha darted toward him, as swift and menacing as a school of hammerhead sharks.

Feebly, Harrison threw a hand up in front of his face to protect himself, as ineffectual as tissue paper held in the path of a heat-seeking missile. He let out a frightened little scream and squeezed his eyes shut, waiting for the cloud of death to squeegee the life out of him.

Mr. Garrote giggled.

Harrison lowered his arm and opened his eyes.

The Alpha floated inches from his face, so close he could reach out and touch it if he temporarily lost his grip on sanity. A reckless part of him wondered what it would feel like. If it would be an agony in every pore and nerve ending in his body or an icy numbness lulling him sweetly into oblivion. Of all the ghosts in the park Mr. Garrote could have sent to kill everyone in the control room, the fact that he'd chosen these faceless assassins, the Alpha to His Omega, spoke for itself.

"*Oh, be a dear and let my pets out, would you, Harry?*" Garrote said. "*They're quite hungry.*"

Harrison did. Gladly.

HARRISON GREELY HAD NEVER BEEN a popular boy.

He couldn't dance, couldn't bring himself to talk to girls without stumbling over his words, couldn't play sports outside of table tennis, couldn't speak up in class without a stutter, couldn't make friends, couldn't impress his teachers with physical prowess or mental aptitude or even gain their sympathy when the other kids picked on him.

What he *could* do was program. He'd been a whiz with computers ever since he stripped his older brother's PlayStation down to its individual diodes and chips at the age of nine, reprogrammed it via their dad's Pentium III, and put all of the pieces back in the correct order. The only difference, other than a slight crack in the gray casing, was that young Harrison had suddenly been able to beat his brother Tommy in any game by using a simple cheat code and still make the win look natural.

Tommy hadn't played with him much after that.

While Harrison was sad to lose his brother's sporadic companionship, the time alone gave him the opportunity to formulate what he'd already begun to think of as his Great Work. With rapid advancements in technology during the early-2000s, old motherboards, casings, RAM and various spare parts were often tossed into dumpsters or left in boxes on front lawns, providing enough supplies to keep Harrison busy between school and supper, when no one was home besides himself and Tommy—and once Tommy left for Penn State in 2004, Harrison alone.

Young Harrison dreamed of supercomputers with enough raw intelligence to subjugate every one of his bullies. Somewhere between his early and late teens the idea of creating an AI Superintelligence began to feel too risky. Nanotechnology could theoretically repair and maintain the Machines just as easily, perhaps easier, than any human. He'd read Kurzweil, von Neumann and Moravec's thoughts on technological singularity and "transhumanism," the theoretical merging of Man and Machine. He'd read about neural networks and "deep learning," which had still been in its infancy when James Cameron's epic techno-fantasy film *Avatar* had been unleashed upon a world Harrison considered woefully unprepared for its genius, breaking every single box office record and forever changing his life, shaping his destiny.

He'd left the theater that night feeling a deep sense of excitement—*my Great Work, at last!*—but also a terrifying urgency. His invention would change the world, the way Cameron's film had changed the definition of what a Blockbuster Film could be, but only if he lived long enough to complete it. He was especially careful on the way home, looking twice before crossing the street, holding his keys in a fist to prevent potential thieves and murderers from harming him, locking the door and returning to check on the locks multiple times throughout the night as he worked alone in semidarkness.

The world would never know they had lost a Great Mind if Harrison Greely died before his Great Work was done. He vowed that someday, the world would know his name. They would fall at his feet and grovel.

In his second year at MIT, a Mormon transhumanist group discovered his work and offered to fund his studies. He'd worked with them throughout the next few years of his PhD, but their partnership had never reached a place of synergy. They were too nice. They'd all wanted eternal youth and happiness. And while Harrison would have been fine with eternal youth, happiness was the furthest thing from his mind.

It wasn't until the Hedgewood Foundation poached him in his last year at MIT that Harrison discovered his true value. They'd offered him a deal: they would pay for his final year of tuition and provide a per diem of fifty dollars a day—far more than Harrison had ever seen in his ramen-and-Gatorade life—in exchange for his Friday evenings and weekends. He'd gladly accepted, not even caring that Hedgewood had a reputation for overworking and underpaying its employees. After all, he wouldn't technically be an employee. They needed his brain more than he needed them, and Harrison certainly didn't need his Fridays and weekends.

His liaison had picked him up that first Friday after class and every subsequent one in a private helicopter. The following Monday, he'd noticed jealous looks from fellow co-eds. Suddenly, people who'd had no interest in him were full of questions. Who was he working for? What was he working on? Could he get them on the team?

All of the employees at Hedgewood, from the janitors to theoretical physicists, ate in the same stark white cafeteria. Non-disclosure agreements prevented them from revealing details about their work to others who weren't on their team, which left most conversations consisting of things happening in the outside world, much of which—since most of these people appeared to live at the facility—was secondhand information from TV.

Did you see what happened on Lost this week? Did you hear what Obama said in his speech? Isn't it awful what happened in Haiti? Can you believe how much oil spilled in the Gulf of Mexico? What about this Julian Assange guy? Is he a hero, a cyber terrorist or what?

Like the sudden keen interest from fellow students back in Cambridge, people at Hedgewood seemed eager to talk to him because of his weekly returns to the outside world. Harrison had stumbled into popularity, yet he still preferred to keep to himself. He listened. He nodded. He didn't have much to offer by way of conversation and when he did, when he'd start to discuss his interests in the Singularity and classic PlayStation vs. next-gen systems, and his favorite movie in the whole world—which had been *unfairly panned* by the critics and *James Cameron was shooting four sequels simultaneously, did you know that? FOUR!*—he would notice their eyes begin to glaze and their attention start to wander, as if they'd suddenly found something extremely interesting to look at on the blank white walls of the cafeteria.

It wasn't long before he fell once more into the role of Pariah, which suited Harrison just fine.

In his third month at Hedgewood, sitting alone at a corner table eating chicken tenders and bright-orange macaroni while the others discussed things he cared absolutely nothing about, Rex Garrote sat down opposite him. In that moment it seemed as if the entire cafeteria had fallen silent.

"I've heard about your work," the writer said—though at the time, Harrison had no clue the man was a writer, nor who he was at all. He was still just a man, not the demigod he would become to Harrison in later years. Harrison had only seen him in passing once or twice, heard people whisper about him secretly, rumors and speculation he felt no need to pry into. Strange things about faking his own death and living here like a hermit since the early 2000s. Stories that couldn't possibly be true—though, in time he would discover they were true, and much worse.

Harrison chewed his food quickly, swallowed hard and choked on it. He washed the blockage down with a mouthful of Gatorade Glacier Freeze. "O-okay," he said finally, not exactly sure how to respond.

The man sitting across from him smiled, and though the smile didn't crease the skin around his eyes it didn't seem to hold a trace of condescension. "You're a man of few words," he said. "I like that. You're a thinker. A *dreamer*, like myself. I see big things in your future, Harrison Greely." He held out his hand. "I'm Rex Garrote."

Harrison wiped his greasy fingers on a stack of napkins and shook the man's cool, dry hand. "Harrison—oh, right. You already..." His skin whispered against Garrote's palm as he let go of it.

"Don't fret, my boy. It's not my intention to poke fun at you or make you feel inadequate."

They sat in silence a moment, Garrote watching him, Harrison feeling scrutinized. Awkward. "There sure are a lot of people here," he said to fill the silence. "Must be a lot of other projects—"

"Not a single one as important as yours." The writer smiled lightly and patted Harrison's hand, which held a forkful of neon macaroni. "Well, perhaps the one other. I'd like to be your partner, Harry. Take you under my wing, so to speak. I feel your work could be incredibly important—could be *great*, in fact."

It was thrilling to hear those words from someone else's lips. That his work could in fact be *Great*.

"You just need a little nudge in the right direction," the writer said. He grinned then: a dark, sardonic half-smile Harrison came to know intimately in the following years. "That, and the proper motivation."

In the control room, Harrison placed the Pandora neural mesh on his head and connected it to the port he'd installed on his terminal without Ms.

Amblin's approval. The emergency lights stopped flashing and the backup lighting flickered on overhead. In the eerie silence that followed he remembered all of the dead who lay here with him. His former colleagues. Some of them he might even have considered friends, in another life. It was a graveyard—both in here and outside.

He'd known all the while that this time would eventually come. That he would stand alone in a room full of death. That he would remain behind, helping a dead man who'd provided him with a sense of purpose he'd known from the moment he'd torn apart his brother's PlayStation would one day be recognized.

This was his Great Work. Right here, right now. The dead were inconsequential.

He called up the Pandora program. On a separate monitor he scanned all active devices, looking for the serial codes Garrote had provided him. On a third he pulled up 3D schematics of the park, showing each of the cameras, every mechanism, door and trick. A fourth showed the main program code, the cursor blinking.

Results pinged back quickly and two separate camera feeds appeared on the device monitor: POVs from the headsets of the boy and the girl. They were running, the images erratic and somewhat blurry. He caught glimpses of the woman they'd come in with, the psychiatrist—as well as two meathead security guards.

A moment later, two flashing red dots appeared on the schematic, very near Bright Falls Sanitarium.

"*There* you are," he breathed. In that moment his grin reflected in the matte black of the coding monitor looked very much like Rex Garrote's own.

PART 1

GHOSTS ARE PEOPLE TOO

It is uncertain exactly what caused the so-called "Ghostland Disaster." Speculation has run the gamut from mass psychosis, to natural disaster, to chemical attack. At this point, it is doubtful the cause will ever be determined, and if it is, whether or not the public will be informed.

War was inevitable, now. Samson felt it in his bones. All he'd ever known was blood. He'd gotten a taste for it soon after he and his brothers set their feet on the shore in Danang, lugging their M-60s, rucks and equipment, greasy sweat sticking their uniforms to their skin. The ghosts of his fallen brothers would fight at his side… and the ghosts of their enemies would stand along with them: Charlie and GIs, shoulder to shoulder, razing the whole godforsaken world to ash.

— Rex Garrote, *Shōki*

THE GIRL WHO PLAYED WITH GHOSTS

Placid Oaks Cemetery
Berkeley, California—October 31st, 2019

"ONLY A PSYCHO would spend Halloween at a graveyard," Lilian Roth said, bundled up against the evening chill among the tombstones. Passing under the soft glow of a cobwebbed streetlamp, Lilian cocked an ear to her left, as if to listen to the autumn wind rustle dead branches. Her laugh was a bright and cheerful counterpoint to the aura of death and decay. "Yeah, you wish, dork."

To most people it might have looked like she was speaking to herself, with no one in the desolate graveyard to hear her.

Lilian wasn't like most people. Her once dull, normal life had been irrevocably changed twice: first, on the day in April when her best friend Ben Laramie and her therapist Dr. Allison Wexler, along with more than two-thousand others, had lost their lives at Ghostland. The second, the night Ben returned to her as a ghost.

He paced alongside her now, mimicking the actions of her arms and legs. The illusion of life wasn't quite perfect: his feet hovered an inch or so above the ground, for starters. Lilian had already gotten used to him occasionally forgetting to breathe or blink or walk, and the fact that he never felt hungry or thirsty or tired. Ben thought it made her uncomfortable when he behaved like any other ethereal being, so he continued to act like the living around her. He'd suddenly puff out his chest as if he was taking a deep breath, or pretend to stumble upon noticing he'd just walked through something at shin-level that would have tripped a living person, and she'd have to try hard not to laugh.

Just a few months ago she never would've imagined she'd be working with Ghosts Are People Too, stalking through a graveyard on Halloween with the ghost of her best friend, in search of a dead woman to free from the woman's own personal Hell. But here she was ascending the low, tree-covered hill in the center of Placid Oaks, where a columbarium stood with a view of the entire cemetery. The mist had grown thick around her feet, and the stone building housing dozens of urns filled with cremated remains shone stark white under the crescent moon. This was where people claimed to have most often seen the so-called "Woman in White."

"Is she even here?" Lilian asked.

"I'm not sure, there's a lot of interference," Ben said, sensing the presence of multiple ethereals lurking among the headstones. "Hey, did you hear about Miss Delyse?" he asked, remembering what he'd heard earlier in the day.

"What about her?"

"She committed suicide."

"*Are you serious?*"

"Dead serious," he said.

It had become one of his favorite phrases, lately. Lilian would've kicked herself for setting up the joke but she was too bothered by what he'd said to even roll her eyes. For a short time after Sara Jane Amblin informed the world about the existence of an afterlife there had been a sharp rise in suicides, especially among teens—but that was over a year ago. The spike had tapered off since, with a drastic drop following the Ghostland Disaster. The world seemed to be slowly returning to normal.

The sweet, middle-aged Caribbean psychic from TV she and Ben used to watch ironically seemed like the furthest thing from suicidal. She was a grandmother. Grandmothers didn't commit suicide.

"That's awful," she said finally.

"News said she like decapitated herself in her car or something," Ben said.

"She cut off *her own head*?"

"Well, like, she tied a rope around her neck to a tree and backed out of her driveway. There were people watching from across the street, they said."

"Why didn't they try to stop her?"

"I dunno, I guess they didn't—" He paused, listening as the wind whistled, rattling bare branches like rolled bones. "She's here."

Lilian startled. "*Miss Delyse?*"

"Jessica, dumb-dumb. Why don't you set up the Ouija on that crypt over there?"

Lilian tugged on the strap of her knapsack, drawing it closer to her body. "I still don't get why we have to use this thing."

"It's the only way anyone's been able to contact her. Like Professor Hermann with that old phone, 'member?"

She remembered Claus Hermann, the bespectacled Jewish scientist originally from Berlin, with his odd little obsessive-compulsive tics and the sporadic outbursts of static electricity that formed around him. They'd only been able to contact him using the old rotary-dial phone in his apartment, which was now part of a historical tenement museum in New York's Lower East Side.

Professor Hermann had witnessed the Third Reich's rise to power in the '30s and fled Germany to America before the situation grew dire. Despite this, he was afraid to join their fight against Rex Garrote—a war Ben and Lilian felt was inevitable, and would reach them before any of them were ready.

Sadly, there were more ethereals like Hermann than Ben and his friends at GRP2. Content with the status quo, repeating the same trivial tasks and motions day after day, hiding themselves from the living who shared their space, often unknowingly. Afraid to fight back, before it was too late for any of them.

"Take me back," Professor Hermann had said after a short time among them. "I prefer my home. *Bitte*."

Ben had taken him back to the museum, where guests would marvel at their hair standing on end and the static shocks he'd unintentionally give them. The next day two others from their group had fled.

"I'm gonna wink out for a bit," Ben said, hovering among the graves. "See if I can shake her loose." Without awaiting Lilian's response, he disappeared from her side.

She still found it sort of weird to see him "wink out," as he called it. The way he acted with her she could almost believe he was still alive, aside from his little mistakes. Then he'd disappear and she'd remember her best friend was the Dead Kid. The kid she'd shunned for so many years to avoid the sting of embarrassment for being his friend.

She sat on the crypt, a cold, dulled-granite block in the vague shape of a coffin with the name UNTERGANG engraved in its side. The plastic planchette rattled inside the box as she heaved her knapsack off her shoulder. She laid the Ouija board on the cold granite beside her, placing the planchette at the center of the board. She lit a candle and laid the tips of her fingers gently on the planchette's plastic, heart-shaped surface. Closing her eyes, she moved it in concentric circles, wider and wider, like a splash in a pond.

It was strange how something so simple could contain so much power, like a cross to a vampire—less magic, Lilian assumed, than psychological conditioning. Like Claus Hermann's telephone. Or the ghost they'd tried to liberate from the supermarket, who would only communicate through condensation on the glass in the frozen food aisle. The Woman in White—also known as the "Singing Woman," due to claims of witnesses hearing the phantom singing of an indistinct song—had only ever responded to the Ouija.

"I wish to contact the spirit of Jessica Kissimon."

She widened her circles, the light scratching of the planchette on the board the only sound aside from the wind in the trees and the swish of dead leaves in the autumn evening chill.

"Jessica Kissimon, the spirit called the Woman in White, I wish to communicate with you! If you are here," she said, projecting her voice, "speak to me through the board. Tell me why you remain in this cemetery. Tell me what you require to move on."

The planchette jittered under her fingers, so lightly it could have been from her shivering. It stopped circling and swept across the board, pausing for a moment with the viewing aperture above the M. It moved to the A and stopped there.

MA.

"Your mother? Did your mother do something to you, Jessica?"

The planchette cycled rapidly between M and A, M and A—*MAMAMAMAMA*. Then it stopped. Lilian held her trembling fingers over it with a creeping sense of dread, her breath caught in her throat.

Jessica Kissimon died in a head-on collision in 2011 on her way back from Vegas, where she and her new husband had eloped. Her husband had been drunk and survived the crash. The family of four in the other car died.

Jessica's mother had been sitting in their living room on the other side of town, watching *America's Got Talent*. She later told the papers she hadn't approved of their marriage but she would have tolerated him if it meant she'd have her baby back.

The planchette moved again, very slowly, dragging across the board, as if some force was pushing against the spirit she'd summoned. It came to rest on the N.

She waited. The planchette held there.

"Man?" she said aloud.

The candle guttered with a low rumble. The clouds parted in the same moment, revealing the stark-white sliver of moon.

"What man, Jessica?"

She covered her ears from the sound of a high-pitched shriek. Stone cracked nearby and she jumped, turning toward it in mid-air. The grave marker directly ahead of her had cracked down the middle. The shriek stopped abruptly. Reluctantly, Lilian lowered her hands, wondering if that was the sound witnesses had spoken about.

"Jessica, I'm a friend," she called out, a nervous quaver in her voice. "I'm trying to help you."

She tried to conjure the calm demeanor Allison had as she'd spoken to the dead man at the farmhouse in Ghostland, drawing him gently out of his body. But she was frightened. No matter how many ghosts she'd encountered since that day, when an ethereal acted out, even out of fear themselves, she couldn't help but worry.

"Jessica? You said man. What man, Jessica?"

The shriek came again, louder this time. Headstones cracked and tipped around her, like the ground splitting on a fault line. The planchette circled, tracing the path of the stones as they toppled one after another. Bare branches whipped and the moon disappeared under a scud of dark clouds.

Lilian drew her knees up to her chest. The second her fingers left the planchette it shot off the board and clattered at the foot of the columbarium steps. The remains of the family of four in the other car were stored in there. The theory was that Jessica haunted the building due to her guilt over their deaths. That she mourned the loss of their lives as much as her own, if not more.

The voice fell silent.

"I've got her!" Ben cried.

He reappeared suddenly, hovering at the top of the steps, holding the Woman in White in a bear hug from behind. She struggled against his grip, the gauzy material of her white, Victorian-looking wedding dress moving languidly, like dancing underwater. Jessica's mother denied her daughter had ever owned such a gown, and it wasn't something she might have picked up for a quickie Vegas wedding. Even the police report had confirmed she hadn't been wearing it when she died.

It made Lilian wonder if Jessica had imagined it into being. Most ethereals wore what they'd been wearing when they died, or what they were buried in. Lilian looked at her own clothes: a pair of loose-fitting gray TNA joggers, one of Blake's bulky Stanford hoodies, her puffy winter vest and her worn Converse. If she died right now and had to wear this outfit for eternity, the afterlife would be pretty embarrassing.

"Jessica, listen to me," Ben said. "And please, *please* don't scream again." He paused a moment, waiting for a reaction. She remained silent, continuing her slow-motion struggle. "I'm like you, okay? Only I never got trapped like you did. I'm not stuck, get it? Don't you ever wonder why you're always wandering around this graveyard night after night? Wouldn't you rather be somewhere else? *Do* something else?"

Her lips moved soundlessly. Lilian thought it looked like someone calling for help in a nightmare. The spirit's eyes widened in absolute terror. A deep red stain began to bloom in the middle of her dress.

"She's trying to say something," Lilian said.

Holding her as tight as he could, Ben tried to look over her shoulder to see if he could make out her words. Her gown fluttered into his field of vision. "Can you tell what she's saying?"

Lilian watched her lips. "It looks like she's saying *Easter*." She squinted, watching the Woman in White repeat the same words as the red stain spread over her breasts and down the length of the gown. "Please pear?"

"'Please pear'?" Ben scoffed. "Yeah, that's probably what she's saying. Totally makes sense."

"Well, I'm not a lip reader! Why don't you Vulcan mind meld her?"

"Because that's an invasion of privacy, Lilian."

"Okay, then just hold her in a bear hug all night. Maybe she'll tire herself out, like, never."

"*Fine.*"

Ben squeezed his eyes shut. His mouth opened and a young woman's voice came from his lips, warped and thick-sounding, like when Lilian had played one of her dad's old 45 records after she'd left the box too close to the radiator.

"*Heeeeeee's heeeeeere,*" Ben said in Jessica's voice.

Lilian felt a chill straight through to her bones. *He's here.* Was "he" the man Jessica warned her about with the Ouija board?

Drained of energy, Ben let his arms drop from around Jessica Kissimon's waist. She immediately vanished in a blinding white light. The columbarium dome cracked. The weathervane tilted, the rooster and arrow issuing a rusty squawk. The final headstone split and the upper half collapsed on the grave mound.

No way could the Woman in White be this powerful.

He's here, Lilian thought. *Garrote? Is that who she means by "he"?*

Ben winked out. For a terrifying moment Lilian worried he'd abandoned her, that he'd run away and left her for dead. He reappeared at her side a moment later. "Come on!"

She leaped off the crypt, knocking the Ouija board aside, and hurried down the hill, the low mist whipping up around her legs. Ben flew ahead of her, quicker than any living person could move.

As they neared the gate, a dark figure stepped out from behind a weathered stone angel. Lilian skidded to a halt, tearing up clods of damp grass. The man was maybe twenty feet from her, two rows away.

The dark figure stepped into the gateway, barring escape. Something in his hand caught a glint of light from the streetlamp beyond the brick wall.

This is it. Nowhere to run. He's come for us. It's all over now.

EXCERPTS FROM KNOW YOUR GHOSTS: A GHOST HUNTER'S GUIDE, COMPILED BY THE GHOST BROTHERS

1 Within these pages you'll find a comprehensive guide to the types of hauntings you'll encounter at GHOSTLAND, the Most Terrifying Place on Earth! We've investigated a bunch of these hauntings in our Documentary Network series *Ghost Brothers* but nothing like the way you'll see them here, "live" and up close.

As expert ghost hunters and fans of pretty much everything paranormal, we had a blast at GHOSTLAND during the creation of this guide and we're sure you'll have a great time here too. So c'mon, guys—let's go check out some ghosts together!

HOW IT WORKS

First, let's run you through the science part really quick, okay? GHOSTLAND features hundreds of exhibits and objects haunted by many types of spectral beings (*see Index*). Each ghost is made visible using a combination of state-of-the-art Augmented Reality and breakthrough Recurrence Field™ technology. Entire buildings have been disassembled and reconstructed here, alongside the most haunted house in America: Garrote House.

With your patented GHOSTLAND AR glasses you'll be able to see things that have never been visible to the human eye before. Our ghosts are actually holograms (or "spectrograms," as we call them here). But they are 100% real, recreated using the most advanced 3D animation available, historical images and data, and actual "dead energy" of once-living beings—all of which are stored on GHOSTLAND's massive server array.

Every single haunting in this park is a digital imprint of the "dead energy" still haunting and/or possessing these buildings and objects, contained by the Recurrence Field™. That means any ghost you see here was once a living, breathing human being—so please, treat them with respect. Like they say at the zoo, please don't tease the "animals"!

GHOSTLAND was the brainchild of deceased horror author Rex Garrote. But the real genius behind the theme park is maverick inventor, Sarah Jane Amblin. If you want to know more about the history of the park, take a tour of the Visitor Center located near the entrance. And if you want to experience the exhibits the way we did, check out our interactive Ghost Hunting Experience in Legion House while you're there!

TYPES OF HAUNTINGS*

ORBS – These are the most basic form of leftover "dead energy," stripped of any human essence. They're essentially harmless and present themselves as white or multicolored balls of light. They're often mistaken for dust in video/photographic evidence. *Also called*: sprites, will-o'-the-wisps. Orbs can be found in many exhibits throughout GHOSTLAND.

REVENANTS – These are full-bodied, floating apparitions. They're potentially harmful when in "solid" form, though they're most often seen in a confused or agitated state. Many examples can be found throughout GHOSTLAND, including Legion House in the Ghost Hunting Experience (Visitor Center), Bright Falls Sanitarium, Historic Beacon Barracks, The Gentlemen of the Sea, Garrote House.

POLTERGEISTS – These are noisy, messy spirits that can move and/or levitate objects using what is believed to be either short-range telekinesis or particle manipulation (further research is being done on this subject at GHOSTLAND's Science and Technology building—which is off-limits to the general public but is a pretty rad place if you're able to get a tour). Poltergeists have been known to physically harm the living but only in minor ways: unexplained scratches, bruises, etc. Prime examples can be found in several Visitors Center exhibits, Fontaine County Correctional, Crane Gardens, Apache Theater.

ELEMENTALS – Similar to poltergeists, though these ghosts can only manipulate water, air, fire, earth and in some cases, electricity or metal. Many vengeful spirits are known to have elemental powers. Prime examples are Mrs. Crane (Crane Gardens), Charles Manafort (The Gentleman of the Sea), the Magnificent Quentin (Midway), the Mime (Merchant Bros. Circus), the Behemoth (Garrote House).

POSSESSORS – These extremely dangerous ghosts can enter living hosts and make them do their bidding for short periods. As with other hauntings able to manipulate objects and living beings, this ability is being researched in the Sci/Tech building. Examples can be found in: Apache Theater, Bright Falls Sanitarium, Starlight Arcade, Garrote House.

DOPPELGANGERS – Appear in the form of someone known to the haunted. Because of this it's been thought they have a low-level ability to read minds, though some paranormal investigators believe they act as mirrors, tapping into the latent psychic abilities of the person being haunted. Found in: Fontaine County Correctional and Garrote House.

TRICKSTERS – Extremely dangerous spirits who use psychological manipulation including many of the above influences to terrorize the living. Though the exhibits at GHOSTLAND have been rendered harmless via the Recurrence Field, it is important to never taunt this type of spirit. Often mistaken for demonic presences. *Also called*: mesmeric ghosts. Found in: Fontaine County Correctional, Bright Falls Sanitarium, Garrote House.

* Most apparitions have very limited abilities to affect their surroundings. Others, like Revenants or Tricksters, can be very dangerous when encountered. It is advisable that Ghost Hunters avoid contact with them at all times, except under the safe conditions at GHOSTLAND.

EXHIBIT INDEX

Welcome To Ghostland

1 REX GARROTE (1947 - 1999) – While not one of our "ghosts," you will encounter Mr. Garrote's likeness throughout the park as a holographic host to many of the exhibits. Mr. Garrote burst onto the horror scene in the late-1970s with *A Roller-Coaster Ride Thru Hell*, a violent, psychedelic retelling of Dante's *Inferno*. He wrote several dozen novels over his lifetime, multiple screenplays and teleplays, and created the short-lived television series *Ghost World*, which he also hosted. His career truly flourished after his purchase of Garrote House, previously the Hedgewood Estate. It was believed by some that the spirits of the house influenced his output from the late-'80s through the '90s. In 1999, his cremated remains were found in his library. Though police suspected self-immolation, the case is still considered "cold." Among all the spirits visible at Garrote House, Mr. Garrote himself has never been

seen.

Know Your Ghosts

1 ROBERT THE ENCHANTED DOLL (*Poltergeist,* 1904) – Thought to be the most haunted doll in the world, Robert originally belonged to Robert Eugene Otto, an eccentric artist from a prominent Key West family. The doll was of German origin, given to Otto as a child by his grandfather in 1904. Some believe the child-sized doll, dressed in Otto's boyhood sailor suit, is possessed by the spirit of a young Bahamian girl. Others believe it was cursed. Robert was said to have moved throughout Otto's house on its own after the artist's death. Footsteps and laughter were often heard in the night, and children claimed to see Robert looking down at them from the upstairs window.

2 JOE "SCHMO" RUSSO (*Apparition,* 1935) – Joe Russo was a small-time hoodlum who joined the ranks of the Oklahoma City mob as an enforcer in the early '30s. Ruthless and violent, Russo's favorite method of inflicting pain was psychological torture. He would give his victims a headstart before chasing them in his car, batting them around like a cat with a mouse. Russo was gunned down by the FBI during a raid. His Model-T was said to have spooked workers at the Henry Ford Museum in Dearborn, MI, with the flashing of its headlamps and the rumble of its engine late at night.

Maniacs

1 GENTLEMAN OF THE SEA (*Various,* 1721) – Charles Manafort was an officer of the British Royal Navy until his ship and crew were held for ransom by Edward Teach (Blackbeard) in 1713. When England failed to negotiate, Manafort and most of his crew agreed to join with Blackbeard—those who did not were promptly executed. Though his exploits were often attributed to Blackbeard, Manafort and the *Gentleman* swept through the Caribbean Sea, mercilessly commandeering lone ships and attacking full armadas. The *Gentelman* was ultimately caught off-guard and surrounded by a French fleet. His crew died in cutthroat combat and he was strung from the mast of his own ship, left to die of exposure, an example to all pirates who dared challenge King Louis XIV. Dry-docked since its capture in Portsmouth, England, one of the only surviving ships of its era, its ghostly crew have been seen manning their posts over the years.

2 APACHE THEATER (*Various,* 1982) – From Rex Garrote's narration: "This sprawling one-story theater was owned and operated by somewhat of a showman who had employed rather ingenious tactics to terrify his audiences. On this particular evening, each of the theater's seats was fitted with a 'shock chair' and wrist shackles for a special one-night-only screening of the cult film House of the Zapper. But their host was a madman bent on murder. He invited only the harshest critics of the film, his personal favorite. When the shock chairs caught fire, he stood in the wings watching until all the critics were crispy. Summing up a career in cinema, when he was condemned to death and given the electric chair his last words were the film's final line, "That's showbiz, folks–and ain't it electrifying?'"

3 DOLL'S HEAD MURDERS (*Poltergeist,* 20) – Six women were murdered over the course of one summer, each victim a young, beautiful woman found nude and heavily made-up, the head of a similar-looking fashion doll placed between their lips. Lead suspect Alexander Robin Fischer, of the esteemed Seattle Fisher family, was released on a technicality. Valencia Motton was the fifth victim, her body discovered in her University Trailer Park home. Her ghost has been known to rattle the door, raise and

lower the blinds, as well as turn on and off the lights despite a lack of electricity.

Stalker

1 THE MAGNIFICENT QUENTIN (*Poltergeists*, 1930) – Quentin Taft was an illusionist of little note. Born an orphan, he spent his youth learning card tricks and illusions to prove his worth to the rich and powerful. He achieved some success between 1899 and 1902, when another illusionist came into prominence. Since then, he remained under Houdini's shadow. His final "illusion" was to drown his lovely assistant Minnie Winslow in the water tank escape while making himself "disappear" by dousing himself with sulfuric acid—to the horror of the unsuspecting crowd.

2 ROCKY'S FUN WORLD (*Apparition*, 1985) – The owner of this traveling midway, one Rocky Arnault, killed himself in a bizarre fashion within his own mirror maze. He had cut pieces off of his body—small at first, then progressively larger and more vital—leaving them at the base of the mirrors at each dead end so that the pieces of himself had lain against their reflections. By the time he'd crawled legless to the stairs leading up to the second level, he had bled out and died, his remaining hand slick with blood, still holding the sawblade. None of Rocky's colleagues or friends could explain this seemingly ritualistic form of suicide and it was later chalked up to the funhouse itself being cursed. Whatever the reason, his funhouse remains the most haunted in all of America, Rocky himself often spotted as a reflection within his hall of mirrors.

Breather

1 GHOST TOWN, USA (*Apparitions*, 1885) – The outlaw Henry Smokes and his gang met the deputy of Lonesome Plains in the desert the night before the massacre and warned him they would be riding through town to rob Great Western Holdings. Sheriff Boone Holden, a self-declared "hero" of the Black Hills War, had never bothered to warn the townspeople, believing he and his deputies could outgun the Buckskin Gang with minimal if any collateral damage. To prove just how wrong the sheriff had been, Smokes had ordered his gang to kill every last man, woman and child in Lonesome Plains before he and the Buckskins rode off with the money. (The Lonesome Plains Massacre is one of GHOSTLAND's main attractions, with shows repeating on the half hour.)

Virtual Insanity

1 MUSEUM OF CURSED OBJECTS (*Various*) – This exhibition currently houses the original Annabelle doll, the Dybbuk Box, "The Hands Resist Him" (painting), the Killing Chair, the Myrtle's Plantation Mirror, and various other high-profile haunted objects from around the world.

2 MORTON WELLES (*Revenant*, 1943) – Also known as the Bright Falls Zombie, Welles was a mild-mannered yet manic patient of Bright Falls Sanitarium made to murder hapless citizens in the surrounding area by his psychiatrist. Dr. Hammersmith—aka Dr. Death; *see*: Bright Falls Sanitarium—considered Welles his "finest work," a "somnambulist [sleepwalker] worthy of the great Dr. Caligari." As Welles's case files were missing, little is known about the former tax collector aside from the murders he committed at the behest of Hammersmith, though it is believed he was self-committed. He

was shot by police at the scene of an attempted murder, and left no surviving family members.

The Garrote Code

1 BRIGHT FALLS SANITARIUM (*Various*, 1882 - 1956) – A "snake pit" asylum in its heyday, during head psychiatrist Dr. Hammersmith's reign of terror, patients were subjected to all manner of psychological and physical torment. Known by the handle Dr. Death among patients and some staff, Hammersmith appeared to have taken a page from Mengele's handbook, though he claimed to have been inspired by the 1920 silent German film, *The Cabinet of Dr. Caligari*. Using a combination of psychic driving and prefrontal lobotomies, Hammersmith created zombified patients who would kill on command. It's been speculated that Hammersmith was on the CIA's payroll, but no evidence has been found linking them. Over the course of his tenure there, at least eighteen patients were murdered with dozens more tortured and abused. The remains of the dead were found among those who had died of tuberculosis.

Survival Instinct

1 CRANE GARDENS (*Various*) – From the *Solitude Fountain* exhibit sign: "The widow Agatha Crane began constructing her garden after the death of her husband, a wealthy Southern plantation owner. This portion of the garden, the yew maze, took an entire year to construct and cost a small fortune. Upon its completion Mistress Crane spent much of her time within this courtyard at the center of the maze, a space she'd named *Solitude*. If a slave disturbed her silence, they were harshly reprimanded. It was rumored that several were never heard from again, believed to have been chased into the maze by Mistress Crane's henchmen, and murdered at her hand." From the *Beauty Fountain* exhibit sign: "The courtyard at the outer edge of the Crane hedge maze, named *Beauty*, is a near exact duplicate of *Solitude* at its center. It was believed for many years that Mistress Crane had ended her life here, drowning herself in this very fountain. Recurrence Field evidence has proven that Mistress Crane did not commit suicide however, but was murdered by the very slaves she had killed."

2 STARLIGHT ARCADE (*Various*) – Opened in 1969 by one Wallace Braugham, the Starlight became a Times Square staple in the '70s and '80s, but fell into disrepair in later years. It was discovered that Braugham was a convicted pedophile from Omaha living under an assumed name, but not before his six victims, boys ranging in age from ten to thirteen, were dredged from the sewers. The interior is a recreation of the actual Times Square arcade, featuring décor and many of the original games, including the *Dr. Dude* pinball machine, the *Angel Knives* and *House of the Dead* games, along with various haunted cabinets from arcades and private collections throughout America.

3 THE DOLLOP HOMESTEAD (*Revenant*, 1947) – The third son of a pig farming family, Jodi Dollop was unnaturally large and is believed to have had mild intellectual disabilities. At a young age he became fascinated with entomology, though it was later discovered he merely enjoyed sticking insects with pins. This behavior progressed to small wild animals and pigs, until one day the Dollops returned from a trip to town to find Jodi's two older brothers impaled on the metal poles Mrs. Dollop used to hang her laundry line. The Dollops did their best to keep the murders secret, but when a traveling vacuum salesman visited the farm while Mrs. Dollop was alone with her son, Jodi murdered her and impaled the salesman with his own merchandise. Since Jodi's death, insects, wild animals and

lost pets have been found spiked and pinned to the walls of the Dollop Homestead.

Dr. Death

1 EMMA LOU AMESBURY (*Trickster*, 1934) – aka Sister Serpent, Emma Amesbury was a nun believed by her convent to have been possessed by Satan. After seducing the priest and found to be carousing with other nuns, she was excommunicated and committed to Bright Falls Sanitarium. While under its care, Dr. Hammersmith allowed her to wear a nun's habit, using role-playing to treat her delusions. She began seducing several members of the male staff—the last of whom she unmanned with her teeth. After her death of tuberculosis, staff claimed to have seen her ghost roaming the halls nude except for her habit. Only one thing seemed to deter her restless ghost: a Christmas snow globe belonging to her former psychiatrist.

2 DR. HAMMERSMITH (*Possessor*, 1950) – aka Dr. Death (*see*: Bright Falls Sanitarium). Hammersmith was put to death for his crimes. His ghost haunts the halls of Bright Falls, particularly the Operating Theater, where he committed the worst of his crimes.

Ghosts In The Machines

1 POLLYANNA (*Poltergeist*) – The Pollyanna doll was an innocent china doll until it was given a new home at the Happy Home for Girls, a Pennsylvania orphanage, in the mid-1920s. There, with the addition of a small child's tricycle, many children claimed to hear "Pollyanna" talking to them, giggling and rolling up and down the halls late at night. Staff claimed the children made up the stories, blaming the doll for their own bad behavior. According to legend, when the strict headmistress was found dead from a third-floor drop, police discovered Pollyanna on her trike in the broken window.

2 CAR 438 (*Apparitions*. 1976) – The Chambers Street crashon the IRT Broadway–Seventh Avenue Line was the worst accident since the 1928 Times Square derailment. In total 141 passengers were injured and six killed, with the majority of deaths in the third car, Car 438, when it struck a wall. A heavily-graffitied Brightliner car, it was decommissioned after the accident and sat in the Concourse Yard for decades. It was said to rock from side to side on the yard tracks and emit a phantom stench of human excrement and body odor, though no evidence of human activity could be found. Rex Garrote's estate spared it from being used as a part of the artificial Redbird Reef off the coast of Slaughter Beach, Delaware among hundreds of other MTA subway cars.

3 SEA DREAM (*Apparitions*, 1987) – New York Stock Exchange broker Buddy Aimes and his mistress, Svetlana Mikhailov, took a cruise on the *Sea Dream* just one day after the Black Monday market crash. Aimes was notorious for selling junk bonds, and lost a fortune. With a poisoned bottle of champagne they toasted his "success." It is unknown whether or not Svetlana was aware of his plan, though Aimes's widow claimed he left her and their two children to be with his mistress. "Good riddance to both of them," she was quoted as saying. The *Sea Dream* coasted for days until it washed up in Bar Harbor, ME, where Aimes had once promised his widow they would buy a summer home once he hit the big time.

4 STUNTMAN LYLE (*Apparition*, 1980) – From Rex Garrote's narration: "Lyle Dabner was one of the most sought-after and death-defying stuntmen in Hollywood during the American New Wave era. He worked with the big names: Scorcese, de Palma, Friedkin, Cassavetes. After a heated argument with

a first-time director over the logistics of a crash, Dabner decided to perform the stunt anyway, driving over the side of the dam while the cameras rolled. The seatbelt rig failed, and Lyle was decapitated on impact. I call this high-octane exhibit, *Stunt Driver Loses Head, Job.*"

Sanctuary

1 FONTAINE COUNTY CORRECTIONAL (*Various*) – Some of the most violent psychopaths in America were imprisoned here at one point or another, from ruthless gangsters to serial killers, rapists and white-collar criminals. The mettle of many hardened criminals was tested in C Block, where prisoners were pitted against each other with whatever weapon they could find, while guards in the "Coliseum's" tower placed bets on the outcome. Fontaine was thought to be a "dumping ground for bad guards." After a riot in 1994 causing the deaths of several inmates and staff, officials decided to act on longstanding advice and close the prison for good. It is considered to be the most haunted place in America, housing the restless spirits of countless men put to death by the state. (Visitors are able to participate in an authentic prison riot experience on the hour.)

Prison Break

1 MR. LIM (*Revenant*, 1878) – Jianguo Lim (Lin) immigrated from China to the Sierra Nevada as a gold prospector in the early 1850s. He formed the largest group of Chinese gold miners in the area, protecting them from attacks from European miners. After years on the panhandle, he lost what little gold he had in a holdup at the Great Western Holdings in Lonesome Plains, Nevada. He worked for several years on the railroads to get back on his feet before purchasing a plot of land in Oregon, where he began his paper mill. By 1878, when he was pushed in front of a train by an unknown assailant, he was one of the wealthiest Chinese landowners in Oregon.

2 STEAMROLLER (*Apparition*, 1995) – A star defensive tackle for the Knee High, Nebraska Bighorns football team, Billy "Steamroller" Becker had a big future ahead of him in the game. While on the road to a championship game against the Huskers he stuck his head out of the bus window. The bus driver had veered too close to the ditch and his head struck a road sign, causing his immediate decapitation.

The House Wakes

1 GARROTE HOUSE (*Various*) – Built by shipping magnate Oliver Hedgewood, Garrote House (formerly the Hedgewood Estate) is not only haunted by the spirits of many former residents, but is believed by some to be plagued by an entity many paranormal investigators and spiritual mediums have called "demonic," or "cursed." While the ghosts you will encounter inside are real, there is no concrete evidence that any such malignant supernatural force exists within its walls. Several deaths occurred during construction of the house, though it was not uncommon at the time due to poor labor conditions. The house was passed down through the generations of Hedgewoods, from Oliver to his widow Charlemagne, to their eldest son Horace, to his eldest Ernest. Upon Ernest's death, the house lay abandoned for several decades, believed by locals to be cursed. Garrote House has a long history of

suicide. Teenagers and indigents fell victim to its allure, and paid the price with their lives. In the late-'70s it was owned by world-renowned sculptor Clayton Odell, who brutally murdered his entire family before killing himself. Finally, it was purchased by Rex Garrote in 1981. The last surviving Hedgewoods sold it for far less than its value. Garrote wrote his most successful books there, and it was suggested he was merely a vessel translating the true horrors of the house onto the blank page. Garrote committed suicide there in 1999, the house abandoned yet again... but never alone.

2 THE BEHEMOTH – Clayton Odell became a darling of the 1970s art scene with his found-object sculptures, fusing human likenesses with mechanical and structural parts. His work was displayed in prominent galleries all over the world, from the MOMA in New York to Kunstmuseum Basel in Switzerland. It wasn't long after purchasing the Hedgewood Estate (Garrote House) that his behavior began seeming "off." Critics and colleagues suspected fame had gone to his head. With less than six months spent in the new house, Clayton sleepwalked to his studio to retrieve some tools, whereupon he killed first his sleeping wife and then his two newborn children. He cut apart their bodies and fused them into a monstrous sculpture made of flesh, which he wore before killing himself. The tragedy made world headlines. However, Odell's art fell into disfavor, sold off to private collectors with a taste for the macabre.

Old Bones

1 OLIVER HEDGEWOOD – A shipping magnate by trade, after Hedgewood built his enormous estate he began to diversify his investments into acquiring curiosities and *objets d'art* of both a macabre and sexual nature from around the globe. His most prized possession was a bronze fist and phallus amulet from Rome, 150- 200 A.D. He was alleged to rub this furiously to achieve arousal. Hedgewood's corpse was discovered in the kitchen by his wife, upon returning from an extended vacation in the Geneva. He had apparently shot himself in the head with a shotgun. He was survived by three children.

For more delicious dark fiction,
visit **www.duncanralston.com**
and **www.shadowworkpublishing.com**.

Want to dive deeper into the world
of *GHOSTLAND*? Visit
www.ghostlandpark.com

Printed in Great Britain
by Amazon

63034265R00182